Air-To-Air Heat Exchangers for Houses

How to bring fresh air into your home and expel
polluted air, while recovering valuable heat

William A. Shurcliff

Brick House Publishing Company
Andover, Massachusetts

Published by Brick House Publishing Co., Inc.
34 Essex Street
Andover, Massachusetts 01810

Cover design: Ned Williams
Typesetting: Janet E. Lendall
Pasteup: Norma Wilton

Printed in the United States of America

Printing: 10, 9, 8, 7, 6, 5, 4, 3, 2, 1
           '85, '84,   '83,   '82,

Library of Congress Cataloging in Publication Data

Shurcliff, William A.
      Air-to-air heat exchangers.

      Bibliography: p.
      Includes index.
      1. Heat exchangers. 2. Dwellings--
Heating and ventilation.   I. Title.
TJ263.S48   697   82-4180
ISBN 0-931790-29-8   AACR2

TABLE OF CONTENTS

INTRODUCTORY CHAPTERS

CHAPTERS ON SPECIFIC COMPANIES' EXCHANGERS

Chapter 1

INTRODUCTION

The increasing interest in exchangers

The coming mandate

Defintion of air-to-air heat-exchanger

Two classes (by function) of air-to-air heat-exchangers

Alternative names

Winter use and summer use

Build or buy?

Big advances expected

Additional information welcomed

Warning

Acknowledgment

Some highly pertinent books

THE INCREASING INTEREST IN EXCHANGERS

Interest in air-to-air heat-exchangers is increasing by leaps and bounds. As recently as 1979 many architects and builders had never heard of them. Yet today practically all architects and builders (and many home-owners also) recognize the need for them in tightly built houses and are anxious to learn more about their performance and cost.

The surge of interest may be explained by these facts:

Our understanding of indoor pollutants and the health-threats they pose has increased.
New kinds of pollutants have appeared on the scene. For example, formaldehyde
    compounds contained in particleboard.
Today's new houses are much tighter than houses built decades ago. Infiltration
    rates have been reduced by factors of 3 to 10. (My friend R. T. Kriebel says:
    "You're caulking your coffin!")
The threat of indoor pollution is being publicized in the popular press. See, for
    example, the article by Zamm and Gannon in the July 1981 Rotarian Magazine;
    it has the frightening title: "Your House Can Make You Sick!"

The dilemma is: how can the home-owner, with his tight house and many sources of indoor pollutants, maintain a health-giving inflow of fresh air and at the same time avoid a big loss of heat in the out-going air? How can he keep the air fresh and the heating bill small?

The answer, of course, is to (1) steer clear of building materials and furnishings that are pollutant-rich, and (2) use an air-to-air heat-exchanger. The exchanger does its job well, may cost only a few hundred dollars, and may have an annual operating cost of only $25 to $60.

2

Why not rely on infiltration -- generous infiltration? Why not make the outer walls etc. leaky enough so that plenty of fresh air will flow in, and install a low-cost solar collector that will make up for the heat-loss? Such a scheme can sometimes work well. But it has limitations. One is that the air-change rate may be too low on calm days and too high on windy days. (In Chapter 16 I describe a radical proposal that might overcome this difficulty.) Another is that, on windy days, an enormous amount of heat is lost with the outgoing air; there is no salvage of heat. Also, in cold cloudy spells the solar collector will provide no help at all.

Why not use electrical air-cleaning devices, such as the ionizing devices that facilitate collection and removal of particulate matter? For example, why not use the "Air Care" ionizing and collecting device marketed by Dev Industries, Inc. 5721 Arapahoe Ave., Boulder, CO 80303? - or the "Modulion" device sold by Life Energy Products Inc., PO Box 75, GPO, Brooklyn, NY 11202? Because these devices (1) are mainly effective with respect to actual particles, whereas a large fraction of room pollutants are gases, (2) do not get rid of excessive moisture, carbon dioxide, radon, etc., and (3) do not replenish the supply of oxygen.

THE COMING MANDATE

Exchangers will soon be a must not only in superinsulated houses but also in many tightly built houses of other type, whether heated by active or passive solar systems or by conventional means (oil, gas, electricity.) The Government of the province of Saskatchewan, Canada, seems to be taking the lead in promoting the use of exchangers: on June 1, 1981, it established a formal requirement that, if the owner of a new energy-efficient house is to qualify for an interest-free loan and if the house has a forced-air-type of heating and ventilating system, this system "...must be designed in such a way that an air-to-air heat-exchanger can be added to the system at a later date, if it is found advisable by the occupant to do so."

Especially large exchangers will be needed in buildings that contain many persons (schools, lecture halls, theatres, crowded office buildings) or contain large sources of pollutants (various factories and chemical plants, for example.)

This book deals mainly with exchangers for houses. Thus it deals with small, simple, inexpensive devices.

DEFINITION OF AIR-TO-AIR HEAT-EXCHANGER

Any device that removes (extracts, recovers) heat from one stream (warm stream) and delivers it to another stream (cold stream) is called a heat-exchanger.

The heat in question may be sensible heat, latent heat, or both. Heat that has raised the temperature of a body is called sensible heat because the rise in temperature can be sensed by a thermometer or by your fingertips. Heat that has changed a material from a solid to a liquid or from a liquid to a gas (without changing the temperature) is called latent heat. Example: a pound of 70°F water vapor contains more latent heat than a pound of 70°F liquid water.

In this book I deal just with exchangers that employ streams of air, i.e., air-to-air heat-exchangers.

TWO CLASSES (BY FUNCTION) OF AIR-TO-AIR HEAT-EXCHANGERS

The two functional classes are:

Air-replacing exchanger Such a device is designed to change the air in a house, in winter, without much loss of heat. The main purpose is to change the air. Saving heat is secondary.

Highly schematic diagram of an air-replacing exchanger

Non-air-replacing exchanger This is designed to supply heat to a house in winter without changing the air. The main purpose is to supply heat. An important requirement is that the initial stream of hot air must not be allowed to enter the house. (Hot air from a paint shop may be too toxic; hot air from a greenhouse may be too humid.)

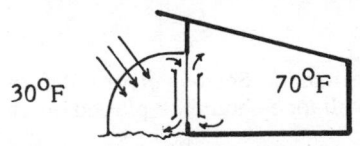

Highly schematic diagram of a non-air-replacing exchanger

This book deals almost entirely with air-changing air-to-air heat-exchangers. For simplicity, I usually refer to them simply as exchangers.

The outgoing air, which may be polluted, smelly, or humid, is often called old air, stale air, or exhaust air. The incoming air, which (in winter) is cold and dry, is called new air, fresh air, or supply air.

ALTERNATIVE NAMES

Sometimes exchangers are called by other names, for example, recuperators. Certain devices that employ sequential flow are called regenerators.

I believe that a new name is needed: a name that is more fully descriptive (implies air exchange as well as heat exchange) and is briefer. I propose the term hexar, for heat exchanger and air replacer. Greg Allen of the Canadian firm of Allen-Drerup-White, Ltd., proposes: heat recovery ventilation unit.

WINTER USE AND SUMMER USE

Most exchangers can be used in winter and also in summer. In winter they save heat. In summer they save "coolth"; that is, they transfer some of the heat in the (hot) incoming air to the stream of (cool) outgoing air, with the result that the fresh air entering the rooms is somewhat cooler than would be the case if no exchanger were used; the need for air conditioning is reduced.

In many parts of the country the winter use is five or ten times more important than the summer use. This book deals almost entirely with winter use.

BUILD OR BUY?

Many kinds of exchangers are commercially available. Some kinds can be homemade. I discuss the commercially available ones at length, and I discuss the homemade types briefly.

4

BIG ADVANCES EXPECTED

The use of exchangers in houses is just starting. Relatively few kinds of exchangers for houses are in routine production. Operating experience is minimal. I predict that the subject will grow fast. I expect many new designs of exchangers to appear in the next two or three years: different designs, improved designs. I expect also that there will be many new technical articles and perhaps several books on the subject.

ADDITIONAL INFORMATION WELCOMED

I will be much obliged to persons who supply me with additional information such as would help me correct errors and omissions in this book or would help me in preparing a later edition. Such information should be sent to William A. Shurcliff, 19 Appleton St., Cambridge, MA 02138.

WARNING

Much of the information presented here has not been fully confirmed. Serious errors of fact, or errors in attributing credit, praise, or blame, may occur. No reliance should be placed on any specific statement without making independent confirmation.

ACKNOWLEDGMENT

I am indebted to many persons for informing me about the design and operation of exchangers. I am indebted to Prof. F.H. Abernathy for general instruction, to J. C. Gray for helpful suggestions, and to David Bearg, J.F. Kreider, H. Kelly, N.B. Saunders, and William Fisk for extensive advice and for spotting bad errors.

SOME HIGHLY PERTINENT BOOKS

Author and title	Bibliography item
Committee on Industrial Ventilation, "Industrial Ventilation: A Manual of Recommended Practice"	C-595
Jakob, Max, "Heat Transfer"	J-110
Kays, W. M. and A. L. London, "Compact Heat Exchangers"	K-52
Knudsen, J. G., and D. L. Katz, "Fluid Dynamics and Heat Transfer"	K-400
McAdams, W. H., "Heat Transmission"	M-86
Sheet Metal and Air Conditioning Contractors Assn., Inc., "Energy Recovery Equipment and Systems: Air-to-Air"	S-162
United States National Bureau of Standards, "Waste Heat Management Guidebook"	U-471m

Chapter 2

AIR: ITS MAIN PROPERTIES

Dry air at standard atmospheric pressure consists of 78.1% (by volume) nitrogen, 21.0% oxygen, and 0.9% of other gases.

The density of ordinary sea-level air at various temperatures -- and at sea-level atmospheric pressure -- is shown in the following table. Also shown are the specific volume, specific heat at constant pressure, and viscosity. (Viscosity is sometimes called absolute viscosity, or dynamic viscosity. A somewhat different concept -- not used in this book -- is kinematic viscosity; it is the ordinary, or absolute, viscosity divided by the density of the gas or liquid in question.)

Some Properties of Air (at Sea Level)

Temp. (°F)	Density		Specific volume (ft^3/lb.)	Specific heat at const. pr. (Btu)/(lb.°F)	Viscosity $\frac{(lb.force)(sec)}{(ft^2)}$
	lb. mass per ft^3	$\frac{(lb.force)(sec^2)}{(ft^4)}$			
25 to 35	0.081	0.0025	12.3	0.24	3.6×10^{-7}
55 to 65	0.075	0.0023	13.3	0.24	3.7×10^{-7}
70	0.074*	0.0023	13.6	0.24	3.8×10^{-7}
95 to 105	0.070	0.0022	14.2	0.24	4.0×10^{-7}

Source: M-86 p. 411,   S-45 p. 165, and other
*Some authors use 0.077. I do too, sometimes (inconsistently!).

Sample use of table:

Question 1: How much heat is needed to raise the temperature of one pound of 69F air to 70F? Answer: Inspection of the table shows the answer to be 0.24 Btu.

Question 2: How much heat is needed to raise the temperature of one cubic foot of air at 69F (and at standard pressure) to 70F? Answer: (0.077 lb.) (0.24 Btu/lb.) = 0.0185 Btu.

Question 3: How much heat is needed to raise the temperature of one cubic foot of air from 30F to 70F? Answer: 40 x 0.0185 Btu = 0.74 Btu. (Here I neglect the fact that 30F air is slightly denser than 70F air.)

Question 4: How much heat is needed to raise the temperature of 10,000 ft^3 of air (about one typical houseful) from 30F to 70F? Answer: About 10,000 x 0.74 Btu = 7400 Btu.

In this table, density is expressed in two kinds of units. The first kind, namely pounds of mass per cubic foot, is most familiar. However, it is the second kind that I use in this book when Reynolds number, turbulence, etc., are under discussion; see Chapter 10.

The specific heat of air changes very little with temperature: less than one part in a thousand for the range from 0°F to 70°F.

6

Moisture Content

The moisture content of air is often of importance, as is made clear in Chapters 4 and 11. To summarize some of the main facts:

70°F air that is at standard pressure and is completely saturated with water contains 0.016 lb. of water per pound of dry air (i.e., per pound of N_2, O_2, etc.) That is, the water content is 1.6%.

Saturated air that is at 30°F contains only 0.004 lb. water per pound of dry air -- i.e., only one fourth as much!

Saturated air at 90°F contains 0.031 lb. water per pound of dry air.

To restate the 70°F situations: If you have 60 lb. of 70°F air that is fully saturated with water, the water content is about 1 lb.

The importance of the moisture is threefold: (1) water vapor contains much energy (latent heat); thus if moist air is discharged to the outdoors there may be considerable waste of energy; (2) if the concentration of water vapor is high, the room occupants may feel some discomfort; (3) condensation of water vapor may cause long-term damage to wood and other materials, may cause frost or ice formation on very cold surfaces, may encourage growth of mold.

Warning I use a variety of terms to refer to the gaseous water that is in air: moisture, water content, water vapor, H_2O. The term "moisture" is short and simple but is not accurate; it has, unfortunately, other connotations.

The latent heat of water vapor is 1054 Btu/lb.

The specific heat of water vapor (at constant standard pressure) is 0.5 Btu/(lb. °F). This is more than twice the specific heat of air.

Chapter 3

THE POLLUTANTS: A BRIEF SURVEY

INTRODUCTION

This is a book about exchangers, not pollutants. Accordingly this chapter is a brief one.

Usually I use the word <u>pollutant</u> rather than <u>contaminant</u>.

Pollutants may be gases, liquid, or solids. They may exist as single molecules or as aggregates (droplets, particles, etc.). They may be defined in terms of chemical composition (radon, nitrous oxide, etc.), in terms of sources (bathroom pollutants, kitchen pollutants, wall-covering pollutants, etc.), or in terms of consequences (smells, carginogens, toxins,etc.)

I call water a pollutant because a high concentration of it in room air can cause discomfort, may produce condensation, may invite the growth of mold.

Water (or moisture, or high humidity) is such an important pollutant that I devote a separate chapter to it. Likewise a separate chapter is devoted to radon.

POLLUTANTS GENERATED WITHIN THE ROOMS

These pollutants include:

Water vapor (H_2O)--from cooking activity, showers, dish washers, clothes dryers,
   occupants' breathing, occupants' perspiration
Radon (Rn), a radioactive gas
Carbon monoxide (CO) -- from gas stove, wood stove, fireplace
Carbon dioxide (CO_2) -- from gas stove, wood stove, fireplace, occupants' breathing
Formaldehyde (HCHO) -- from plywood, particleboard, adhesives, insulation,
   furniture padding
Other aldehydes (molecules containing the group $-\overset{\text{H}}{\underset{|}{\text{C}}}{=}\text{O}$)
Chemical sprays, e.g., for killing flying or crawling insects
Particulates (respirable suspended particles) -- from general dust, smoke from
   gas stove or wood stove, smoke from cigarettes, etc. The diameters of pertinent
   particulates range from 0.001 to 100μ m. Wood stoves produce some "polycyclic
   organic materials" (POMs) some of which may be carcinogenic.

8

Less important materials: usually only negligible quantities of the following materials are generated within the rooms:

Nitrogen monoxide (NO)
Nitrogen dioxide (NO_2)
Sulfur dioxide (SO_2)
Ozone (O_3)
Asbestos
Lead (Pb)
Water soluble nitrates (molecules containing the group NO_3)
Water soluble sulfates (molecules containing the group SO_4)

POLLUTANTS FLOWING UPWARD INTO THE ROOMS

The most significant such pollutants are:

Water and water vapor. If the water-table is high and the basement floor is wet (or is permeable), much trouble may arise. See Chap. 4.
Radon (Rn). Many investigations have indicated that much of the radioactive gas radon may accumulate in tightly sealed houses that are built on granitic terrain, or employ concrete basement walls, or employ brick walls, or masonry Trombe thermal-storage walls, or contain a bin-of-stones thermal storage system.

POLLUTANTS ENTERING FROM OUTDOORS

Country houses and city houses that have leaky roofs and are situated in regions having much rainfall may be plagued with excessive moisture.

Country houses may be polluted by suspended organic matter (spores, e.g.) that enters from outside. Sprays used to improve garden crops, trees, etc. may pose problems.

City houses may be polluted by a variety of chemicals produced by nearby automobile and truck traffic, factories, restaurants, etc.

Such pollutants may enter a house via open doors or windows, or via air intakes for furnace, fireplace, etc., or via general infiltration.

Sometimes the concentrations of pollutants in outdoor locations is greater than concentrations indoors. An air exchanger can actually increase the indoor concentrations of such materials.

WHAT FACTORS MAKE THE POLLUTANT CONCENTRATIONS HIGH?

Obviously, the concentrations will be high if:

The natural rate of infiltration is very low, because
  a) The house is protected from the wind -- protected by nearby hills, buildings, woods
  b) The general construction of the house is very tight: walls, roof, and basement are tight. This may be the case if

Much use is made of large, low-permeance, snugly installed plates of Thermax, Styrofoam, Thermoply, or the like in the external walls
Much use is made of vapor barriers; no holes have been cut in them; edges are overlapped and sealed
There are few external doors and windows, and such doors and windows have been caulked or weatherstripped
Entrances are of vestibule (air-lock) type
External doors and windows are seldom opened

c) There are no chimneys. Or there are chimneys but they have been equipped
   with tight-fitting dampers.
d) Vents intended to serve the kitchen and bathroom are kept shut

There are no blowers or equivalent for expelling stale air and bringing in fresh air.

The rate of indoor generation of pollutants (or introduction of pollutants) is high.

WHAT HARM DO THE POLLUTANTS DO? WHAT CONCENTRATIONS ARE HARMFUL?

In general, clear and accurate answers are not yet available. Information is fragmentary.

Bad smells	Usually these are relatively harmless. Annoying, yes. But, usually, they produce little demonstrable harm.
Humidity	Troublesome if too high or too low. See Chap. 4.
N_2O, NO, NO_2	These gases can harm lung tissue. Permissible indoor concentrations have not yet been finally established. (0-145)
Formaldehyde vapor	Concentrations exceeding $100 \mu g/m^3$ cause irritation to eyes and upper respiratory systems. (F-700). In the Netherlands a limit of $120 \mu g/m^3$ has been proposed (F-700). Some proposed upper limits on indoor concentrations are: California and Wisconsin: 200 ppb; Denmark and Netherlands: 100 to 120 ppb. (0-145).
CO	50 ppm 8-hr. average is OSHA occupation standard. 30 ppm is 1-hr. National Ambient Air Quality Standards limit.
CO_2	500 ppm for 8-hrs exposure indicated by ASHRAE. (F-700)
O_3	0.08 ppm. (F-700)
Radon	See Chap. 5

MEASURING THE CONCENTRATIONS

Methods used at the University of California are described in Reference 0-145. Other methods
are described in papers given at the October 1981 conference discussed in a later chapter.

Badges Responsive to Formaldehyde

Small "badges" that can be attached to a person's clothing and will measure one-week cumulative
exposure to formaldehyde have been produced and put to use. I understand that duPont Co.
markets such a badge, called Colorometric Pro-Tek Air Monitoring Badge System, Series II, Type
C-60.

STRATEGIES FOR REDUCING THE CONCENTRATIONS

Among the available strategies are:

"Forbid" the concentration. Prohibit smoking. Prohibit cooking cabbage. Prohibit use
   of sprays. Prohibit long showers.
Remove the source of pollution. Transfer the source to an outdoor shed or city dump.
Block off the polluting object. For example, cover the earth floor of a crawlspace
   with a plastic sheet impervious to radon and moisture.

Provide local ventilation for an especially bothersome source.
Provide general ventilation, as by means of an exchanger.
Employ electrostatic precipitators or air filters (of fibrous or charcoal type.)
Open windows and doors.

Of course, persons building and furnishing houses should try to pick a site where little radon is present in the earth, use building materials that are free of pollutants, and select furnishings that are non-polluting.

SOURCES OF INFORMATION

There was a major conference on indoor air pollution on October 13 - 16, 1981, at Amherst, Mass. Organized by Harvard University public health experts and government experts, the conference was called "International Symposium on Indoor Air Pollution, Health and Energy Conservation." A proceedings volume (I-430) is to be published, probably in mid-1982. About 100 papers were presented on nearly all aspects of indoor pollution, including: the pollutants, their abundance, their sources, their harmfulness, monitoring methods, and air-to-air heat-exchangers.

A comprehensive report on indoor pollutants was published later in 1981 by the National Academy of Sciences, 2100 Constitution Ave., Washington, DC 20418. Prepared by the Academy's Committee on Indoor Pollutants, the report contains much material on radon and other pollutants. See Bibl. item N-25.

The National Academy of Sciences has published, also in 1981, a comprehensive report (340 pages) on formaldehyde as a pollutant. See Bibl. item N-24m.

Chapter 4

WATER AND HUMIDITY: PRELIMINARY CONSIDERATIONS

Introduction

Basic definitions

Occupant comfort

Durability of materials

Saving latent heat

INTRODUCTION

To have at least a moderate amount of water vapor in room air is highly desirable. If there were no moisture here, occupants would feel much discomfort. Also, objects of wood and various other materials would shrink and perhaps crack and physically deteriorate in other ways.

But a large amount of moisture may be harmful in many ways. The occupants may experience discomfort. Water droplets and frost may form on cold surfaces; small puddles may form, wooden objects may rot, and steel may rust. Growth of molds may proceed rapidly. When (in winter) room air is expelled to outdoors, an unnecessarily large amount of energy may be lost with it -- the latent heat of the water vapor.

In summary, control of humidity of room air is important from three viewpoints: occupant comfort, durability of materials, and energy conservation. These topics are discussed in detail below. But first, some basic definitions are presented.

BASIC DEFINITIONS

Humidity

The humidity (water content) of room air can be expressed in two ways:

Absolute humidity. This is the amount of water per unit amount of air. Amount is expressed in terms of mass, i.e., pounds of mass. Usually "per unit amount of air" is taken to mean "per unit amount of dry air," i.e., per pound of the air constituents other than water. Typical values of absolute humidity of 70°F indoor air are 0.004 to 0.012 lb. water per lb. dry air.

Relative humidity (RH). This is the ratio of the actual absolute amount to the maximum possible absolute amount for air of the given pressure and temperature. Typical values, for room air, are 0.25 to 0.75, i.e., 25% to 75% RH.

Example Consider the (70°F) air in my living room. If the absolute humidity of this air is 0.0078 lb. water per lb. dry air, the relative humidity is 50%.

Dry-bulb and wet-bulb temperatures

The ordinary temperature of air (temperature measured with an ordinary (dry) thermometer) is called the dry-bulb temperature, or merely temperature.

If the bulb of a thermometer is kept moist (as by wrapping the bulb in a small piece of thin cloth that is wet) and the thermometer is waved through the room air at about 15 ft./sec., the thermometer reading will be lower than if the bulb had been left dry -- because of the cooling effect of the evaporation of some water from the wet bulb. The depressed reading is called the wet-bulb temperature, or wet-bulb reading. Example: consider standard-pressure air that is at 70°F and 50% RH; its wet-bulb reading is 58°F.

Saturation temperature (dew-point temperature, condensation temperature)

This is the temperature to which a quantity of air (of given pressure, temperature, and absolute humidity) must be cooled in order to initiate condensation of gaseous water (in the air) into liquid water. (During the cooling process, the absolute humidity remains unchanged until the saturation temperature has been reached and condensation has started.)

Example Consider standard-pressure air that is at 70°F and has an absolute humidity of 0.0078 lb. water per lb. dry air (corresponding to 50% RH): the saturation temperature of such air is 49°F. Note: this is much lower than the wet-bulb temperature.

Latent heat of gaseous water in air

This is the heat that is released when one pound of gaseous water in a large quantity of air is condensed into liquid water, or, conversely, is the energy that was supplied at some time in the past to evaporate one pound of water into this air. It is assumed that, during such process, the temperature of the air is held constant or that correction is made for any energy input or output associated with any temperature change.

The value of the latent heat of gaseous water is: 1054 Btu per pound of water. This value is broadly applicable: it applies throughout a wide range of air pressure and temperature.

Comment: Water "prefers to be liquid, not gaseous." The forces between molecules tend to draw them together. To pull molecules apart from one another, to form a gas, takes a lot of energy. Atmospheric pressure is almost irrelevant to this topic. Even if you arrange to evaporate some liquid water into vacuum, much energy is required -- about 97% of the above-mentioned 1054 Btu/lb. Evaporating the water into atmospheric-pressure air (entailing "pushing the air aside to make room") accounts for only about 3% of the 1054 Btu/lb.

OCCUPANT COMFORT

In indoor 70°F air, relative humidity levels much higher than 60% can be oppressive. Such levels in 75F air can make persons feel uncomfortably hot, especially if they are used to cold climates and are exercising strenuously.

Under special circumstances it is possible to remain comfortable even when the humidity far exceeds 60%. Natural or artificial (fan-driven) breezes can make such levels tolerable.

Even 100% humidity may be tolerable if there is a strong breeze and if the air temperature is well below skin temperature. If the skin is warmer than the air, some water will evaporate from the skin even if the humidity of the room air is 100%. (I first learned of this at a talk given by B. Givoni at the 5'th National Passive Solar Conference, at Philadelphia, in June 1981.)

DURABILITY OF MATERIALS

The colder air becomes, the less water vapor it can retain. When warm, high-RH air comes in contact with a cold surface, some of the water vapor in this air will condense and form droplets (dew) on the cold surface. Consider, for example, 70F air that has a relative humidity (RH) of 80% or greater. When such air comes in contact with a surface that is as cold as 63F or colder, condensation of water vapor occurs. More generally, if a quantity of 70F air has a relative humidity at least as great as 80%, 60%, 40%, or 20%, condensation will occur on surfaces that are at least as cold as 63F, 55F, 42.5F, or 25F respectively. For 65F air and RH values of 80%, 60%, 40%, and 20%, the condensation temperatures are 59F, 49F, 39F, and 22F. For 60F air and RH 80%, 60%, 40%, and 20% the condensation temperatures are 54F, 45F, 34F, and 18F.

If wooden objects in rooms (wooden window frames, sills, floors, etc.) are kept at temperatures higher than the above-listed condensation temperatures, no condensation will occur and rotting may be avoided. Likewise, steel objects will not become damp and may not rust.

SAVING LATENT HEAT

Clearly, if, in winter, much humid air is discharged from a house to the outdoors much energy is wasted.

To compute the magnitude of the waste, one must know how much air is discharged and how much moisture it contains, i.e., the concentration of moisture. Likewise one must know the properties of the outdoor air that replaces the discharged air.

Much energy can be saved if an air-to-air heat-exchanger is used. In discussing the performance of the exchanger, further considerations of air temperature, humidity, etc., arise. Also, questions as to condensation of water arise. Calculations of energy gains and losses are needed.

All of these topics are discussed in detail in Chapter 11.

Chapter 5

RADON: AN INSIDIOUS POLLUTANT

Introduction

Radon: what is it?

Origin of radon

Half-lives of the commonest radon nuclei

What is the radon disintegration process?

Energies of the disintegration products

The physical damage they do

Radon daughters

Unit of amount of radioactivity

How do the radon and radon daughters enter the indoor air?

Can the influx of radon from the earth beneath the building be stopped?

How does the indoor radon get into the human body?

Relative harmfulness of radon and radon daughters

Time delay before the harm becomes evident

Concentrations of radon in outdoor air at ground level

Concentrations of radon and radon daughters in houses

Limit on permissible concentration in houses

Overall harm to persons in USA

History of the recognition of the harmfulness of radon

Some unanswered questions concerning permissible amount

Some major conferences and reports

INTRODUCTION

Radon is such an important and insidious pollutant that I devote an entire chapter to it. Many articles on the radon threat have been published but are far too brief.

An excellent discussion of the subject is contained in the book Radiation Protection by Prof. Jacob Shapiro of the Harvard School of Public Health, Harvard University Press, Cambridge, MA 02138. Mid-1981. 500 p. Hardbound. $25. I have been fortunate in receiving much help (in writing this chapter) from Dr. Stephen Rudnick of the Harvard School of Public Health. See also the late-1981 book by J.W. Gofman: Radiation and Human Health, Sierra Club Books, 530 Bush St., San Francisco, CA 94108. 910 p. $29.95.

RADON: WHAT IS IT?

Radon is one of the 92 naturally occurring elements. It is one of the heaviest ones. The radon nucleus contains 86 protons, and accordingly radon has been assigned the atomic number 86. It is called $_{86}$Rn.

Radon contains a much larger number of neutrons: about 134 of them. The naturally occurring radon nuclei contain 133, 134, or 136 of them. Thus the <u>total</u> number of particles (total number of nucleons) in such nucleus is 219, 220, or 222. These three types (three isotopes, or nuclides) are called ^{219}Rn, ^{220}Rn, and ^{222}Rn.

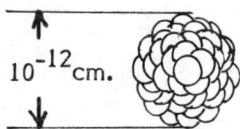

10^{-12}cm.   Radon nucleus. It contains 86 protons and almost twice as many neutrons.

Every radon atom has an enveloping cloud, or set of shells, of electrons. The outermost shell is of special type, called full, or <u>closed</u>, and accordingly the atom has practically no external electrical effects (as long as the nucleus remains quiescent!). It ignores all other atoms of every type whatsoever, and exists in splendid isolation, i.e., as a monatomic chemically inert gas ("noble gas").

Radon nuclei are unstable. Eventually every such nucleus disintegrates, or "explodes". There is no known way to prevent this and no known way to control or predict when the disintegration of any given atom will occur.

Consequently if you have in your house a large quantity of radon atoms -- say a billion of them -- you will find that, every little while (at random times) one atom will disintegrate. Thus radon is said to be radioactive.

Before a week has gone by, more than half of the initially present radon atoms will no longer exist: more than half of them will have disintegrated.

But if your house is built of materials that contain radium, thorium, uranium, or certain other heavy elements, or if the underlying earth contains such materials, radon is being produced all the time and is entering the air in your house.

ORIGIN OF RADON

The radon nuclei that exist today in the air and in the ground were not always here. They were not here when the earth was formed..

They have been <u>created</u> here -- created relatively recently. They are continually being created from heavier nuclei, especially uranium and thorium. The radon nuclei are produced indirectly: uranium and thorium atoms disintegrate and some of their disintegration products disintegrate in turn and produce radon nuclei. Specifically:

Uranium produces a chain of disintegrations. One product is the radium-226 nucleus

which in turn disintegrates into ^{222}Rn.

Thorium produces a chain of disintegrations. One product is the radium-224 nucleus

which in turn disintegrates into ^{220}Rn.

(Less important is actinium, which produces radium-223 and ^{219}Rn.)

The commonest types of radon nuclei are ^{222}Rn and ^{220}Rn.

HALF-LIVES OF THE COMMONEST RADON NUCLEI

The half-life of ^{222}Rn is 4 days. If you have a large quantity of the radon at a given time, about half of it will be gone (will no longer exist) after 4 days.

The half-life of ^{220}Rn is 1 minute.

These are short half-lives! Things happen fast. Hence the threat to health of persons exposed to radon. (If the half-lives were a million years, say, as is true of certain other kinds of radioactive nuclei, disintegrations would be so infrequent as to pose only a very small threat.)

WHAT IS THE RADON DISINTEGRATION PROCESS ?

What, exactly, happens when a radon nucleus disintegrates? The radon nucleus itself ceases to exist and (within a trillionth of a second) these three new particles spring into being:

One slightly lighter nucleus (polonium nucleus),
One much lighter nucleus (helium nucleus, also called alpha particle),
One bit of electromagnetic energy (one photon of gamma radiation).

The following diagrams show how the two main radon nuclei disintegrate.

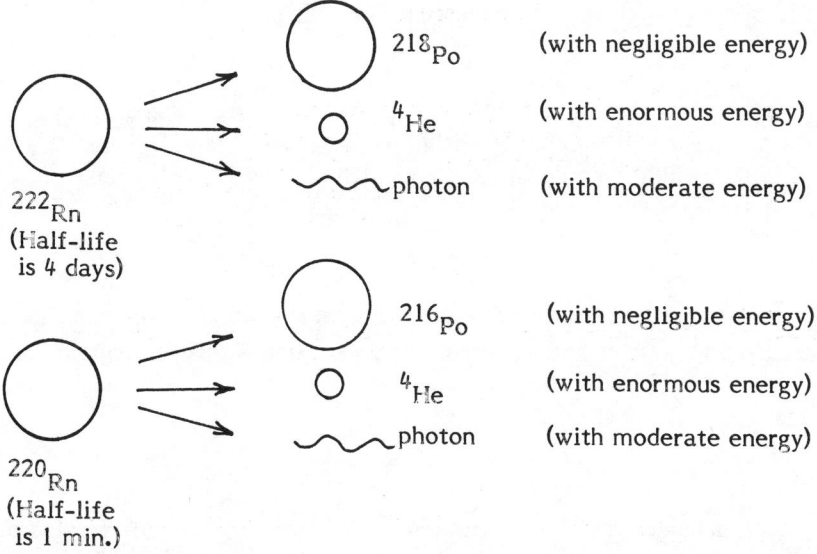

ENERGIES OF THE DISINTEGRATION PRODUCTS

The lion's share of the energy released in the disintegration process is carried by the helium nucleus (i.e., alpha particle). At the instant of its creation it has about 6 million electron volts (Mev) of energy. This energy has the form of kinetic energy: the particle is traveling at very high speed.

A much smaller share is carried by the photon: about 1 Mev. The photon travels, of course, at the speed of light: 3×10^8 m/sec. Only a trivial amount of kinetic energy is carried by the (heavy!) polonium nucleus.

THE PHYSICAL DAMAGE THEY DO

Helium Nucleus

Tremendously disrupting effects are produced by the helium nucleus -- in the matter that the nucleus travels through; more exactly, they are produced in the (solid or liquid) matter that lies within about 0.001 inch of the path taken by the nucleus.

Why are the disrupting effects Large? Because:

(1) The nucleus has an electric charge: a double charge (written ++).
(Why does it have an electric charge? Because the nucleus travels "stripped", i.e., without its pair of electrons; accordingly the ++ charge of the nucleus is not neutralized.)

(2) The nucleus is traveling at extremely high speed. A charged particle that is traveling at such speed does not swerve appreciably when it approaches a charged particle (positively charged nucleus or negatively charged electron); it may come extremely close to such particle and thus give it a large "jolt". Traveling in human tissue, for example, the fast, charged particle may knock some electrons out of some of the molecules, thus changing their chemical nature. Normal molecules (in living cells, for example) may be converted into abnormal molecules.

(3) The nucleus travels a large distance (say, 1/10 inch), producing disruptions all along the way -- about one million disruptions, for example.

Photon

The photon produces a smaller, but still important, amount of disruption. The amount is smaller because the photon has no electric charge and accordingly loses energy only slowly, i.e., in well-separated encounters. Only a slight amount of damage is done in any 1/10-inch segment of path. In other words the physical effects produced by the photon are spread out and have only minor significance at any one spot.

Polonium Nucleus

The immediate damage done by the polonium nucleus is negligible -- because this nucleus, being very heavy, has little speed, little kinetic energy. (Of course, much damage is done by the polonium nucleus later, i.e., when it disintegrates.)

RADON DAUGHTERS

By "radon daughters" (or radon decay products) one means a succession of nuclei formed by a succession of disintegrations. Typically there is a chain, or series, of disintegrations, with about four members to the series. Thus the set might more properly be called radon's daughters, grand-daughters, great-granddaughters, and great-great-granddaughters. Because there are several types of radon, there are many kinds of daughters -- about a dozen in all. All are massive (only slightly lighter than radon itself) and all disintegrate energetically. Three disintegrate especially energetically. All of the radon daughters have atomic numbers of 83, 84, or thereabouts.

Are the chains, or series, extremely long? No. Each stops when the resulting heavy nucleus is a stable form of lead. (Lead is the 82nd element, and the lead nucleus ^{206}Pb is stable.)

What are the half-lives of the main radon daughters? About 10 or 100 minutes, typically. Some of the half-lives are 10 or 100 times shorter, and some are much longer.

The four most potentially harmful radon daughters are ^{214}Bi, ^{214}Pb, ^{214}Po, and ^{218}Po.

UNIT AMOUNT OF RADIOACTIVITY

The curie (abbreviated Ci) is the usual unit of measurement of the radioactivity of a large quantity of radon, radon daughters, or other radioactive material. It is defined as that quantity of nuclei of the specified material -- radon, for example -- that provides 3.7×10^{10} disintegrations per second.

Notice that the type, or violence, of the disintegrations is not taken into account -- as far as this particular unit is concerned. Accordingly the unit is not a unit of energy release, or damage to material, or harm to living matter. Thus the significance, or usefulness, of the unit is limited.

But the unit has this in its favor: it can be cleanly defined, and it clears the way for a simple and accurate method of measurement. Counting of disintegrations is easily accomplished with a variety of low-cost instruments, e.g., Geiger counters, scintillation counters.

Usually, amounts of radioactivity are specified in terms of pico-curies (pCi). One pCi = 10^{-12}Ci. It corresponds to about 2 disintegrations per minute.

HOW DO THE RADON AND RADON DAUGHTERS ENTER THE INDOOR AIR?

They enter partly from the earth (soil, rocks, groundwater, etc.) that is close beneath the house, and they enter partly from concrete or masonry floors, walls, fireplaces, chimneys, etc. There are other means of entry also. Let us discuss the various means of entry separately.

Radon From The Earth

The process starts with uranium and thorium in the ground. These materials produce, over the course of millions of years, radium. The radium produces radon. Although the uranium and thorium (in granite, for example) are relatively immobile, the radon can move easily because it is a gas; it dissolves in groundwater and drifts along with it. Eventually the radon may reach the surface of the earth -- surface of the crawlspace of a building, for example. Thus much radon may get into the building from below.

Radon From Concrete And Masonry

Such materials (especially in parts of the world where uranium is abundant) often contain uranium and radium, and consequently produce some radon, some of which diffuses into the air.

Radon From Ambient Air

A little radon enters a house along with the infiltrating ambient air. This radon may come from many miles away. (But not much comes from air that has been at high altitude, because air that is at high altitude usually stays at high altitude for many months and, by that time, most of the included radon atoms will have disintegrated.)

Radon "Born" Within The Human Body

Some radon is "born" inside the human body. A typical person contains 3×10^{-11} gram of radium, which disintegrates (with very long half-life) into many radioactive products, including radon.

Radon Daughters "Born" In Indoor Air

Some radon atoms that are in indoor air disintegrate there, producing "born indoors" radon daughters.

CAN THE INFLUX OF RADON FROM THE EARTH BENEATH THE BUILDING BE STOPPED?

Yes. It can be stopped by installing an air-tight barrier (an ordinary vapor barrier, say) at the surface of the crawlspace earth. Or by providing a concrete basement floor. The point is, the radon has a half-life of only a few days; if its trip from earth to indoor air is slowed sufficiently that the trip takes several weeks, nearly all of the radon will have disintegrated before it reaches the indoor air. That is, merely slowing the radon's travel may suffice.

Why do not the disintegration products of radon carry on and reach the indoor air? The most important disintegration products are elements that are normally solid, hence not highly mobile. They are not gases, and in particular they are not noble gases; thus they tend to stay put.

Of course, the provision of a concrete floor will not help much if the concrete itself contains much uranium and is itself producing radon.

HOW DOES THE INDOOR RADON GET INTO THE HUMAN BODY?

Most of it enters by inhalation: it enters when we breathe. It enters the windpipe, then the bronchi, then the microscopic pockets (alveoli) which have walls only a few molecules thick (to allow oyygen to pass through into the bloodstream and allow CO_2 to pass in the opposite direction).

Because the radon is soluble in water, it may remain in the human body for a long time. Some, of course, may be promptly exhaled.

Some of the radon will disintegrate while in the human body, and the resulting heavy radio-active atoms may lodge firmly within the body, doing considerable harm as they in turn disintegrate.

Radon Daughter Atoms

Such atoms -- in room air -- tend to adhere to dust particles, and when these particles are inhaled and in turn adhere to the walls of the bronchi, trouble looms. The radon daughter atoms will remain in the bronchi for a relatively long time -- a time long enough for the atoms to disintegrate and do harm Some radon daughter atoms do not adhere to dust particles, and it is believed that these atoms, if inhaled, may be outstandingly harmful

Small amounts of radon daughter atoms may enter the body via ingestion. Some of the food we eat contains traces of uranium, radium, radon, and radon daughters, and these may become incorporated in the body.

RELATIVE HARMFULNESS OF RADON AND RADON DAUGHTERS

It is now known (S-146, p. 172) that the harmfulness of radon daughters (lodged in the bronchus of the lung) is 5 to 500 times the harmfulness of radon itself. The exact factor depends on just which part of the bronchus is under consideration and on the extent to which the bronchus is able to get rid of such radioactive atoms. There are several lung-cleaning mechanisms, for persons who do not smoke. For smokers, clearance is very slow -- fatally slow in some instances. (Clearance is slow enough so that it is not of much help with respect to particles that disintegrate within a few days. This applies to smokers and non-smokers.)

TIME DELAY BEFORE THE HARM BECOMES EVIDENT

Clear symptons of harm (from deposition of radon daughters in the lung) may be absent for many years. Harm may show up, often, 10 to 40 years later.

CONCENTRATIONS OF RADON IN OUTDOOR AIR AT GROUND LEVEL

These values are reported by Shapiro (S-146):

Location	Concentration of radon at ground-level locations	
	pCi/liter	pCi/m^3
Over large continents	0.1	100
Over coastal areas	0.01	10
Over oceans and arctic regions	0.001	1

Note: the values given are for ^{222}Rn only. If ^{220}Rn were included, the values would be slightly higher. The values do not take radon daughters into account.

The global rate at which ^{222}Rn is emitted from land areas into the atmosphere is 50 Ci/sec. This corresponds to a rate of 0.4 pCi per second per square meter of land area.

The total amount of ^{222}Rn in the atmosphere at any given moment is 25,000,000 Ci.

CONCENTRATIONS OF RADON AND RADON DAUGHTERS IN HOUSES

Shapiro (S-146) reports radon concentrations of:

   0.01 to 0.2 pCi/liter in a brick apartment house in Boston, with a fresh-air
      input rate of 5 to 9 house-volumes per hour.
   26 pCi/liter in an unventilated bedroom (in Chicago) over an unpaved crawl-
      space.

Representative value for the average concentration of ^{222}Rn in indoor air: 1 pCi/liter, according to Shapiro (S-146, p. 364).

LIMIT ON PERMISSIBLE CONCENTRATION IN HOUSES

In August of 1981 there were no formal, binding, federal or state regulations as to permissible concentration of radon and/or radon daughters in houses in USA.

There were, however, some strong recommendations.

The recommended permissible concentrations are expressed in terms of Working Level. This unit may be defined as follows: suppose that there is, in each liter of air in a house, at noon today, enough of the important (short lived) radon daughter atoms to produce, cumulatively and ultimately, enough alpha particles to deliver 130,000 Mev of energy. Then the concentration of radon daughters (in this house at noon today) is said to be one Working Level (WL).

Recommendation

The main recommendation is that, in houses in USA, the concentration of radon daughters shall not exceed 0.01 or 0.02 WL -- and that at the very worst it shall not exceed 0.05 WL. (Obviously, such recommendation is not simple and businesslike. Presumably any firm standard announced will be fully businesslike.) Recommendations such as are stated above may be found in the US Federal Register 44 38664 1979 in a document prepared by the US Environmental Protection Agency, and may be found also in a report issued by the US Surgeon General. The recommendations were drawn up in connection with problems arising in certain parts of Colorado and Florida where dangerously high levels of indoor radioactivity had been found. Somewhat similar limits have been proposed by the Canadian government.

Persons routinely working 170 hours a month at jobs involving exposure to radon and radon daughters are permitted to experience higher concentrations during their working hours: about 0.3 WL (time averaged). Thus each year they may receive, while at work, up to 12 x 0.3 WL = about 4 Working Level Months. (A Working Level Month is the exposure, or dose, received by a persons exposed for 170 hours to 1 WL.) Note that the level pertinent to working hours -- 0.3 WL -- is 15 times the above-mentioned 0.02 WL level pertinent to the general public.

If the limit 0.02 WL is to be respected, how does the rate of fresh-air infiltration affect the limit on concentration of radon? Three cases must be considered:

Case 1. House that has a typical amount of infiltration
   Here a general assumption is made to the effect that about half of the radon daughter atoms play no part: they are not in the house (they escaped, or were never formed within the house in the first place) or, if they are in the house, they plate out onto (become permanently attached to) the walls, ceiling, etc.; thus they have no chance of entering the lungs of any occupant of the house. The other half of the radon daughter atoms remain in room air -- free or attached to very tiny particles -- and may be inhaled and may lodge in the bronchi or lungs. The fraction of the atoms in question that remain in the room air, free or attached to particles, (more exactly, the fraction of the energy deposition capability associated with the alpha particles from these atoms) is called the equilibrium factor. In the present case (house with typical amount of infiltration) the equilibrium factor is 0.5 and the 0.02 WL limit corresponds to (makes permissible) a radon concentration of about 4 picocuries per liter.

Case 2. <u>House that is very tightly built</u>
Here an equilibrium factor of about 0.75 may be assumed. Thus to keep
within the 0.02 WL limit, one must restrict the radon concentration to
about 3 pCi/liter.

Case 3. <u>The house is ideally tightly built and none of the radon daughters plate out:</u>
<u>all remain airborn and can be inhaled.</u>
Here the equilibrium factor is about 1.0. Then 0.02 WL corresponds to
about 2 pCi/liter.

These conclusions may be paraphrased thus: Concentrations should be kept below 4, 3,
and 2 pCi/liter for houses that are typically tight, very tight, and ideally tight-and-non-plating.

<u>Note:</u> A person spending a full year in a house in which the concentration of ^{222}Rn is 1 pCi/liter
will receive a whole-lung dose of 600 mrem and a basal-cells-of-the-bronchial-epithelium dose
of 3200 mrem, "..assuming an equilibrium factor of 0.5", according to Shapiro. (The unit <u>mrem</u>,
or milli-roentgen-equivalent-man, is usually used when the entire human body is exposed to a given
amount of radiation. The unit is such that about 500,000 mrem has a 50-50 chance of killing a person.
A dose of 5000 mrem per year has often been considered acceptable for radiation workers. A dose
of 600 mrem just to the whole-lung -- not to the entire body -- is thus very small compared to the
dose considered acceptable for radiation workers.)

OVERALL HARM TO PERSONS IN USA

It has been estimated (by C.D. Hollowell of Lawrence Berkeley Laboratory, per a recent article
by Stephen Budiansky in <u>Environmental Science and Technology</u>) that indoor exposure to radon and
radon daughters may be producing 1000 to 20,000 cases of death-from-lung-cancer each year in USA.

HISTORY OF THE RECOGNITION OF THE HARMFULNESS OF RADON

Radon's harmfulness was first suspected in 1930, when persons working in radium mines in Europe
were found to contract many cases of cancer.

Calculations made in 1940 indicated that, in those mines, radon alone was not sufficiently
abundant to cause so many cancer cases.

In 1951 came the realization that the miners were inhaling -- not only radon -- but also radon
daughters, and the concentrations of the latter in the lungs far exceeded the radon concentration there.
The daughter products, after having remained for a while suspended in room air and after having been
inhaled by the miners, tended to become trapped in the bronchi and lungs. Thus the radon daughters
did far more harm than the radon itself did.

In 1979 the Environmental Protection Agency published <u>recommended</u> limits on the levels
of radon and radon daughter concentrations in indoor air. See <u>Federal Register</u> 44 38664 1979.

SOME UNANSWERED QUESTIONS CONCERNING PERMISSIBLE AMOUNT

Should there be just one published value of <u>permissible amount,</u> or should there be different values
for different parts of the country, different kinds of terrain, different climates (very cold; not cold),
different kinds of houses, different ages of house occupants?

Is the published <u>permissible amount</u> in line with other threats to health? Should it be applied
even to those persons who habitually smoke, drink, overeat, drive recklessly, etc? Should it be applied
to persons who are too poor to keep their houses warm in winter, too poor to buy air-to-air heat-
exchangers? Should it be applied in an era when there is a real threat of a nuclear war?

SOME MAJOR CONFERENCES AND REPORTS

An international two-day conference on "Radon and Radon Progeny Measurements" was held in Montgomery, Alabama, on Aug. 27 - 28, 1981. Most of the 30 papers dealt with methods of identifying and measuring the concentrations of radon and radon daughters and typical concentrations found.

At the Oct. 1981 conference "International Symposium on Indoor Air Pollution", at Amherst Mass., about 15 papers on radon were presented. For details concerning the Proceedings volume, see Bibliography item I-430, or send inquiries to Prof. John D. Spengler, Harvard School of Public Health, 665 Huntington Ave., Boston, MA 02115.

Late in 1981 a very important report was issued by the Committee on Indoor Pollutants, a committee established by the National Academy of Sciences, 2100 Constitution Ave., Washington, D.C. 20418. The report, containing much information on radon, is priced at $16.25. 560 p. Paperback. (Bibl. N-25).

The late-1981 900-p. book by J.W. Gofman "Radiation & Human Health" (Sierra Club Books), $29.95, contains a detailed and highly readable account of radon daughters and their threat to health.

Chapter 6

NATURAL INFILTRATION

Introduction

Causes

Flowrates

INTRODUCTION

In every house there is some natural inflow of fresh air. No house is hermetic. Fresh air may enter via cracks in foundations, walls, window frames, roof, etc. Also it may enter via cracks at tops, bottom, and sides of outside doors. It may enter via chimneys associated with furnace, cooking stove, wood stove. It may enter via vents associated with kitchen or bathroom air-exhaust systems, or via vents intended mainly for use in summer. Occupants may leave windows open. Children may leave outside doors open.

The subject has been investigated by many engineers here and abroad, and there are many interesting reports on the subject. Some are available through Air Infiltration Center, Old Bracknell Lane, Bracknell, Berkshire, Great Britain RG 12 4AH. An excellent report by A. K. Blomsterberg and D. T. Harrje: "Approaches to Evaluation of Air Infiltration Energy Losses in Buildings" has been published in ASHRAE Transactions 85 (1979) Part 1.

CAUSES

Natural infiltration is a consequence of wind and chimney effect (stack effect). These provide the driving forces.

Wind, or the pressure produced by it, drives air into a house on the upwind side of it. The air enters via cracks, holes, etc. If the airflow through a hole is purely turbulent (say because the linear speed of flow is so high), the volume rate of flow, in cubic feet of air per minute, is proportional to the square root of the pressure. If the linear speed is low and the diameter of the hole is small, the flow may be laminar, and in this case the volume rate of flow is directly proportional to the pressure. (B-330). Ordinarily, the type of flow is between pure turbulent and pure laminar and the relationship between volume airflow and pressure is between square root and linear.

Chimney effect (tendency of warm air to rise) is important on very cold days. As warm air leaves via the upper parts of the house, cold air is drawn into the house at lower locations, especially lower upwind locations.

FLOWRATES

The amount of infiltration may vary enormously from day to day and even hour to hour, depending on speed and direction of wind, extent to which windows and vents are left open, and other factors. Cracks may be wider during very dry periods and narrower during moist periods inasmuch as wood tends to give off or take up H_2O during very dry or very humid periods -- and has a slightly greater volume when it contains much H_2O.

Of course, the amount of exfiltration equals the amount of infiltration. Otherwise the house would threaten to explode or implode! Usually it suffices to talk just about the infiltration.

What is the typical rate of inflow of fresh air via infiltration? According to rumor, the rate for typical houses that are at least 10 or 20 years old is about 1 to 10 house volumes per hour, depending on the size of the house, the type of construction (wood? brick? concrete?), care used in construction, type of sheathing used, condition of windows and doors, extent to which the house is exposed to high winds, and various habits of the occupants. For houses 5 to 10 years old, the rate is usually lower: say ½ to 5 house volumes per hour. For houses being built today by contractors concerned about saving energy, the rate is probably even lower: about 1/3 to 2 house volumes per hour. For certain very carefully built superinsulated houses the rate may be 1/20 to 1/3 house volume per hour.

The rates are seldom known -- seldom known even roughly.

In any event, it is usually true that no single number can be satisfactory -- partly because the rate changes from day to day and hour to hour, and partly because each room may have a different rate (especially if doors between rooms are kept shut).

It seems to be deplorable that so many architects and engineers often ignore these facts -- often speak as though a given house had a definite rate of natural infiltration, such as "...1 air change per hour", when, in fact, the rate can be anywhere from 0.1 to 3, say, depending on outdoor temperature and windspeed. The variations are important! They may imply serious hazard to health, and they certainly imply occasional enormous waste of heat.

Chapter 7

FORCED AIR CHANGE: TERMINOLOGY OF OVERALL PERFORMANCE

Introduction

Terminology of air classes

Parameters pertinent to the exchanger itself

Parameters pertinent to the air in the house

Parameters pertinent to the combination of exchanger and house

Important distinction between relative input rate and replacivity

Relationship between replacivity and unit relative input rate when mixing is continuous

Replacance vs. rate of air input: general relationship when mixing is continuous

INTRODUCTION

In any full description of an air-to-air heat-exchanger in use in a house, or any full account of the performance, many parameters are involved. Some pertain just to the exchanger, some just to the house, and some to the combination of exchanger and house.

Here we define and discuss the main parameters. Some of them are tricky; they must be defined with care and used with care.

TERMINOLOGY OF AIR CLASSES

Confusion is avoided if the terminology of air classes is clear. I use these terms:

Old air:	Those molecules that are in the house now and were in the house at some specified earlier instant, such as the instant when an exchanger was turned on.
New air:	Those molecules that are in the house now but were not there at the specified earlier instant. They have been introduced since that instant.
Room air:	The air in the room. It may be old, partly old and partly new, or entirely new. Usually I assume it to be at 70°F.
Incoming air: (Fresh air)	The air traveling (via the exchanger) from outdoors to indoors. Usually I assume it to be relatively pollution-free and to be much colder than room air.
Outgoing air: (Stale air)	The air traveling (via the exchanger) from indoors to outdoors.

Warning

The definitions are tricky! For example, room air is not synonymous with old air, but, often, is a combination of old air and new air. Likewise outgoing air (stale air) is not synonymous with old air, but, often, is a combination of old air and new air.

Assumption as to non-reentrance of air: I assume that, always, the two outdoor ports (for incoming air and outgoing air) are far enough apart so that no discharged molecules find their way into the stream of incoming air. (No corresponding assumption is made for the indoor ports: obviously, much mixing will occur indoors and a few newly arrived molecules quickly find their way into the stream of outgoing air.)

PARAMETERS PERTINENT TO THE EXCHANGER ITSELF

Fresh-air Flowrate

This is the amount of air (in ft^3/min.) driven by the exchanger-incoming-air blower. (Ordinarily, this is the same as the amount driven by the exchanger-outgoing-air blower.) The symbol used here is R_{ex}. R stands for rate and ex stands for exchanger.
Why not use mass flowrate, e.g., lb/min? No good reason except that most engineers are more familiar with volume flowrate. It is a little easier to estimate the volume of air in a house than the mass of that air.

Exchanger Efficiency (more exactly: sensible-heat exchange efficiency)

This is the ratio of (1) the amount of sensible heat that is recovered, thanks to the exchanger, from the outgoing air and transferred to the incoming air, to (2) the greatest conceivable amount that could be recovered by an exchanger. (Here I assume that the heat-exchange sheets are impermeable to water.) The symbol used here is E_E. Many authors use eta: η.

Example If the indoor air is at 70°F and the outdoor air is at 30°F, and if the flowrates in the two streams are equal, and if the incoming air is at 60°F as it enters the room,

$$E_E = \frac{60 - 30}{70 - 30} = 0.75.$$

More generally, the sensible-heat exchanger efficiency may be defined thus:

$$E_E = \frac{T_4 - T_2}{T_1 - T_2} \qquad \text{or the equivalent} \qquad \frac{T_1 - T_3}{T_1 - T_2}$$

where T_1 is the temperature of the (warm) air in the room,

T_2 is the temperature of the (cold) outdoor air,

T_3 is the temperature of the outgoing air as it reaches outdoors,

T_4 is the temperature of the incoming air as it reaches the room.

(These are the symbols used by Reay (R-25).)

Note Re Efficiency Vs. Effectiveness

Many writers call E_E the effectiveness. They do not use the word efficiency. I decline to go along with this, mainly because effectiveness has so many connotations; it might mean effectiveness in replacing stale air with fresh air; it might mean effectiveness in reducing the concentrations of radon, humidity, etc.; it might mean cost-effectiveness. The word efficiency seems to me to be far more apt. It applies, of course, just to the heat-exchange process. It has nothing to do with electrical or mechanical design of the blowers used, nothing to do with the power required by the blowers, and nothing to do with any "COP" of the combination of blowers and exchanger. Most writers occasionally use the word efficiency -- deliberately or inadvertently; I use it routinely, deliberately!

Efficiency Of Exchangers Involving Recovery of Sensible Heat; Extent To Which Water In The Outgoing Air Is Recovered

These topics are somewhat complicated and are discussed in Chap. 11.

Power Use

This is simply the amount of electrical power (kWh) used to run the exchanger blowers and other powered components if any. If additional amounts of power are used, as for defrosting the exchanger periodically, or providing supplementary heat to the incoming air, these should be specified separately.

PARAMETERS PERTINENT TO THE AIR IN THE HOUSE

Volume Of Air In House

Given the plans of a house, anyone can compute the volume of air in it quite easily.

However, a question may arise as to whether to include the volume of air in the basement, crawlspace, attic, vestibules. Are these to be considered regular parts of the house? Is air-change necessary here too? Or are these regions mainly external to the living regions? There is no easy answer.

I call the house-volume V_H, where H stands for house.

a typical 1000 ft^2 house has a volume of about 8000 ft^3.

Concentrations of Pollutants

See Chapters 3, 4, and 5.

PARAMETERS PERTINENT TO THE COMBINATION OF EXCHANGER AND HOUSE

Here I consider parameters that depend both on the properties of the exchanger and on the properties of the house. Defining these parameters takes a little care. Reasons for downgrading and renaming the main parameter used by other authors (air changes per hour) are explained in a later paragraph.

Relative Input Rate

This is the quantity (volume) of air, in terms of the unit <u>one house volume</u>, delivered to the house by the exchanger per hour. The symbol for <u>one house volume</u> is V_H. The symbol for <u>relative input rate</u> is R_i, where <u>R</u> stands for rate and <u>i</u> stands for input.

<u>Example</u> If 20,000 ft^3 of outdoor air is injected each hour into a 10,000 ft^3 house, then

$$\text{Relative input rate} = R_i = \frac{20,000 \text{ ft}^3/\text{hr}}{10,000 \text{ ft}^3/V_H} = 2\ V_H/\text{hr}.$$

<u>Another example</u> If 30,000 ft^3 is injected into a 10,000 ft^3 house in 6 hours, then

$$\text{Relative input rate} = R_i = \frac{\left(\dfrac{30,000 \text{ ft}^3}{6 \text{ hr}}\right)}{10,000 \text{ ft}^3/V_H} = 0.5\ V_H/\text{hr}.$$

Replacance

This is the fraction of air molecules, in the house at some specified instant, that were <u>not</u> in the house at an (earlier) reference instant. In other words, it is the fraction of the molecules (indoors, at a specified instant) that are new, relative to the reference instant.

Note that the definition is <u>not</u> phrased in terms of how many new molecules <u>have been</u> introduced -- because some of those may already have been expelled. The crucial quantity is how many new molecules <u>are in the house now</u>. Because there is continual mixing of the air within the house, some of the recently introduced molecules have already been expelled.

<u>Example</u> The exchanger in Smith's house has been running for many weeks. Tests made at 3:00 p.m. today showed that, at that time, (3:00 p.m.), 90% of the molecules in the house were new since noon. Thus the replacance, at 3:00 p.m. relative to noon, was 90%.

Replacivity

This is the replacance at one hour after a specified reference time, or starting instant. Thus it is a conventionalized, or standardized, replacance: the replacance after one hour.

<u>Example</u> A small exchanger is installed in a large house and is started up at 10:30. One hour later (at 11:30) the replacance is 40%. Accordingly the replacivity likewise is 40%.

<u>Another example</u> In a certain 30-minute running period an exchanger produces a replacance of 50%. What is the replacivity? Answer: 75%. Why? Because if 50% of the old molecules remain after 30 minutes, the fraction remaining after one hour would be (0.50)x(0.50) = 0.25. Thus the replacance after one hour would be (1 - 0.25) = 0.75 = 75%.

Note: These statements are correct only if the newly introduced air is continually and vigorously mixed with the old air. Usually one assumes that such mixing does occur. (Often it does not, as explained on a later page.)

IMPORTANT DISTINCTION BETWEEN RELATIVE INPUT RATE AND REPLACIVITY

These two parameters are very different -- not even proportional. In some situations the relative input rate may be very large while the replacivity is very small. Or vice versa.

Example of high relative input rate, low replacivity Consider a house that has many partitions and many closed doors between rooms. Suppose that an exchanger providing high relative input rate is installed in one small room. Then the molecules of air in this particular room may be replaced with new ones at very frequent intervals, but in many other rooms the air may remain old for many hours or days (because of the partitions and closed doors). For the house as a whole the replacivity is very small.

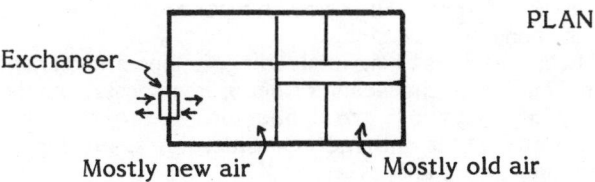

PLAN

Exchanger

Mostly new air   Mostly old air

Another example A certain house is equipped with an average-size exchanger, but the exchanger's two indoor ports are so close together that much of the incoming air is promptly caught up in the stream of air that is about to be discharged to outdoors. Most of the new molecules are quickly expelled. Most of the old molecules remain. In such case the fresh-air input rate might be, say, 3 V_H/hr (very high!) yet the replacivity might be, say, 15% (very low!).

In some special situations, the replacivity may be surprisingly high. Consider a long slender house with no partitions and practically no internal air currents other than those induced by the exchanger. Suppose that the incoming air is injected at the east end of the house and the old air exits at the west end. Then if the fresh-air input rate is 1 V_H/hr the replacivity may be, say, 0.9. The point is, the incoming air progressively sweeps the old air ahead of it, toward the exit port. Much old air is expelled. Almost no new air is expelled (except after a considerable delay -- delay corresponding to the time taken for the new air to travel the length of the house).

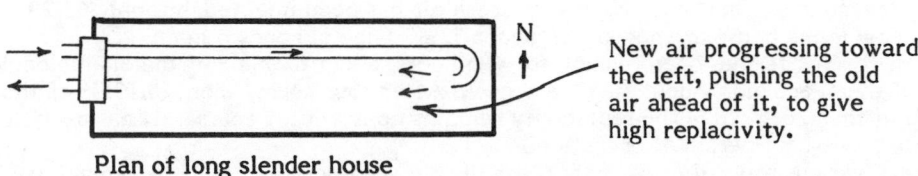

N

New air progressing toward the left, pushing the old air ahead of it, to give high replacivity.

Plan of long slender house

Importance Of Rate Of Mixing Of Newly Introduced Air

If the exchanger's indoor ports (for inlet and outlet) are close together, which is sometimes the case unfortunately, it is highly desirable that there be vigorous and widespread mixing of the air in the room. The faster the mixing, the better, because of (1) reduction in amount of new air that immediately becomes caught up in the air that is about to be ejected, and (2) reduction in stagnant pockets of old air in remote corners of the room in question, or in a remote room. (Such mixing is harmful if, as in the case sketched above, the house is long and narrow and unobstructed and new air is injected at one end and old air leaves via the other end.)

Some Warnings

The literature of air-to-air heat-exchangers is very weak as regards the topic discussed above. It has these faults:

It fails to emphasize the distinction between relative input rate and replacity.

It deals almost exclusively with relative input rate -- and overlooks the fact that the purpose of the exchanger is to provide high replacity.

In dealing with the (non-crucial) parameter relative input rate, it uses an inept and misleading term: air-change rate, or air changes per hour. The reader has the impression that the topic under discussion is replacing old air with new, whereas in fact the topic often is merely the amount of fresh air delivered by the exchanger -- irrespective of whether this air is used in strategic manner or unstrategic manner. In my opinion a grave mistake is made if one uses the words air change when one means air input. The two concepts are different. Even when air input rate is high, air change can be small.

The expression "air changes per hour", besides being misleading, is long and cumbersome.

The usual abbreviation "ACH" (meaning "air changes per hour"), is offensive partly because a reader might think that A, C, and H are to be multiplied together and partly because the important word "per" is omitted. Consider how confusing it would be if "miles per gallon" (mpg) were abbreviated "MG."

It is important that the quantity that really counts -- replacity -- be brought into the fore-front. I don't particularly like the word, but I haven't been able to think of a better one.

RELATIONSHIP BETWEEN REPLACIVITY AND UNIT RELATIVE INPUT RATE WHEN MIXING IS CONTINUOUS

The Relationship

Consider a simple one-room house that is so tightly built that there is no natural infiltration. Suppose that an air-to-air heat-exchanger is continually bringing in fresh air at the rate of one house-volume per hour (1 V_H/hr.) and driving out old air at that same rate. Suppose, furthermore, that at all times the incoming air is thoroughly mixed with the air in the room, so that, at any one instant, the proportion of new-to-old air is the same throughout the room.

Then the fact is that the replacity is about 2/3. More exactly, it is: $(1 - \frac{1}{e}) = (1 - 0.368)$ = 0.632. (The letter e stands for the well-known mathematical constant 2.7183...^{e} which is the base of the natural logarithms.) One house volume of fresh air has been injected, but only 63.2% of the air molecules now in the house are new ones: about 37% of the old ones remain.

After the exchanger has been running for two hours, about $(0.368)^2$ of the old molecules, i.e., about 13.5% of them, are old. About 86.5% are new. After four hours, about $(0.135)^2$ of the molecules, i.e., about 2% of them, are old. (The replacity remains constant, of course: it has been defined with respect to a one-hour period of operation.)

There are various ways of proving that, for the above-described case, the replacity is 0.632. Below I present rough proofs, then a true, accurate proof.

Very Rough, Two-Step Proof

Suppose that we suddenly inject into the one-room house a half-roomful of new air (and, or course, eject a half-roomful of old air) and then thoroughly mix the new air with the remaining old air. The resulting mixture is 50% new, 50% old. Now let us suddenly inject another half-roomful of new air (and eject an equal quantity of the 50-50 mixture). Notice that the first ejection process ejected no new air, but the second ejection process ejected a quarter-roomful of new air -- because the air as a whole was then a 50-50 mixture and we ejected half of it. In summary we have injected one entire roomful of new air and, now, 75% of the molecules in the room are new and 25% are old. Thus the replacity is 0.75, for this two-step process.

Slightly Better (Four-Step) Proof

Let us inject the new air in four quantities (four steps) instead of two: that is, let us inject a succession of quarter-roomfuls instead of two half-roomfuls. As before, we assume that no mixing occurs while a sudden injection is taking place and that thorough mixing occurs after each such injection. Then, in the four injection processes, these volumes of old air are ejected:

1st process: 1/4 x 1 =	0.2500 roomful
2nd process: 1/4 x (1 - 0.25), that is, 1/4 x 0.75, or	0.1875 roomful
3rd process: 1/4 x (1 - 0.25 - 0.1875), that is,	0.1406 roomful
4th process: 1/4 x (1 - 0.25 - 0.1875 - 0.1406), that is,	<u>0.1055 roomful</u>
Total:	0.6836 roomful

In summary, 68.36% of the old molecules are gone. Correspondingly 68.36% of the molecules now in the room are new. The replacivity (for the four-step process) is 68.36%.

The chain of reasoning used in the four-step process can be summarized thus:

$$\text{Replacivity} = 1 - (1 - ¼)^4 = 1 - (0.75)^4 = 1 - 0.3164 = 68.36\%.$$

True, Accurate Proof

This proof requires some use of higher mathematics. One assumes that a series of steps is used. One repeats the entire process again and again, each time using a larger number of steps. The steps become smaller and smaller, and there are more and more of them. The result is a little different each time, and it approaches a limit as N, the number of steps, increases without limit. Accordingly one must find the limit of

$$1 - (1 - \frac{1}{N})^N \text{ as N increases without limit.}$$

The limit is a definite number, well known to mathematicians, and is called (1 - 1/e), where e is the well-known constant 2.7183... The limit is (1 - 1/(2.7183)) = (1 - 0.368) = 0.632 or 63.2%.

Graphical Representation

The following graph shows how the concentration of old air decreases as fresh air is introduced (with continuous ideal mixing) at the rate of one house-volume per hour. The curve is a simple exponential decay curve.

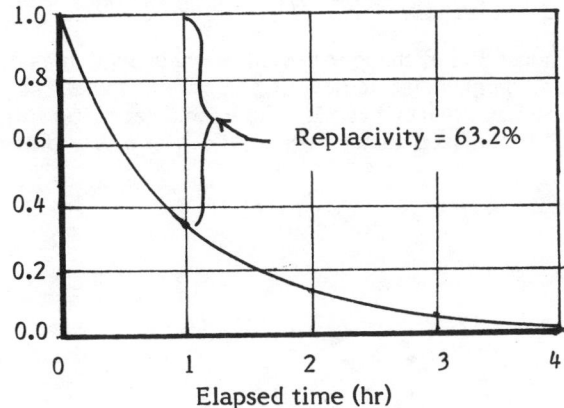

Fraction of old air remaining, as a function of time, if fresh air is injected continually at the rate of 1 V_H/hr. with steady mixing

REPLACANCE VS. RATE OF AIR INPUT: GENERAL RELATIONSHIP WHEN MIXING IS CONTINUOUS

Knowing the relationship discussed above (namely, that with continuous and vigorous mixing, unit rate of air input per hour provides a replacivity of 63.2%), anyone can quickly find the relationship for any other rate of input.

For example, if the rate of air input is 2 house volumes per hour, then the input in ½ hour is one house volume and the replacance at the end of this ½ hour is 63.2%, that is, $(1 - 0.368)$. At the end of one hour the replacance is $(1 - (0.368)^2)$, or $1 - 0.1354$, or 86.5%. Thus the replacivity is 86.5%.

If the rate of air input is 4 house volumes per hour, the input in ¼ hour is one house volume and the replacance at the end of ¼ hour is $(1 - 0.368)$. At the end of the hour the replacance is $1 - 0.368^4 = 0.982$. Thus the replacivity is 98.2%.

Suppose the rate of air input is only ½ house volume per hour. How does one find the replacance at the end of one hour? A square root is required. The answer is $1 - 0.368^{½}$ or $1 - 0.606$ or 0.393. Thus the replacivity is 39.3%.

The general formula for the replacance at the end of X hours as a function of the fresh air input rate of N house volumes per hour is:

$$\text{Replacance} = 1 - 0.368^{XN}.$$

The following table shows replacance values for nine values of time-from-start-of-run and nine values of rate-of-air-change. Replacance is expressed as percent.

Relationship Between Replacance and Rate of Air Input for a Wide Range of Rates Of Air Input, When Mixing is Continuous and Vigorous

Rate of Air Input (House Volumes/hr) →		Replacance (percent)								
		1/16	1/8	1/4	1/2	1	2	4	8	16
Time from	1/16	0.4	0.8	1.6	3.1	6.1	11.7	22.1	39.3	63.2
Start of	1/8	0.8	1.6	3.1	6.1	11.7	22.1	39.3	63.2	86.5
Run (hr.)	1/4	1.6	3.1	6.1	11.7	22.1	39.3	63.2	86.5	98.2
	1/2	3.1	6.1	11.7	22.1	39.3	63.2	86.5	98.2	99.9
Replacivity Values →	1	6.1	11.7	22.1	39.3	63.2	86.5	98.2	99.9	100
	2	11.7	22.1	39.3	63.2	86.5	98.2	99.9	100	100
	4	22.1	39.3	63.2	86.5	98.2	99.9	100	100	100
	8	39.3	63.2	86.5	98.2	99.9	100	100	100	100
	16	63.2	86.5	98.2	99.9	100	100	100	100	100

Example In Smith's house the exchanger provides 4 house volumes of fresh air per hour. What percent of the air in his house is new at 12:30 p.m. if the test run started at noon? Noticing that the time from start of run is ½ hour, and referring to the table (see 7th column, 4th row of the tabulated data on replacance), one sees that the replacance in question is 86.5%.

Chapter 8

HOW DOES POLLUTANT CONCENTRATION DEPEND ON RATE OF AIR INPUT?

Introduction

Case 1: No further introduction of pollutant

Case 2: Continuous introduction of pollutant

Effect of natural infiltration

INTRODUCTION

The main purpose of an exchanger is to reduce the concentrations of pollutants in the rooms.

To eliminate the pollutants is impossible, usually. The concentrations decrease exponentially and thus in theory never quite reach zero.

But to greatly reduce them is simple enough: install an air-to-air heat-exchanger. Using certain simple formulas, one can compute how fast the concentrations will decrease after start-up of the exchanger.

Formulas are needed for two cases:

Case 1: No further introduction of pollutant,

Case 2: Continuous introduction of pollutant.

In each case I assume that there is no significant amount of natural infiltration. If there were such, (a) reliable calculations of pollutant concentration, or decrease in concentration, could not be made, (b) the concentration, or decrease in concentration, would vary from time to time, depending on wind and chimney effect, (c) unnecessary loss of heat would occur, and (d) the need for the exchanger would be less.

CASE 1: NO FURTHER INTRODUCTION OF POLLUTANT

Here I assume that at some specified instant (noon, say) the concentration of a specified pollutant is known; call it C_i, meaning initial concentration. I assume (1) that this pollutant has been thoroughly mixed with the air in the house, (2) it does not decompose or "plate out" onto room walls etc., and (3) no more of this pollutant is introduced.

The question is: How does the concentration C of this pollutant change when an exchanger is put into steady operation (with continuous vigorous mixing of room air)?

The answer is: the concentration is proportional to one minus the replacance. That is, at any time after noon, $C = C_i (1 - \text{replacance})$.

This is logical enough -- because the fate of the pollutant is exactly parallel to the fate of the old air. When X percent of the old air has been eliminated, X percent of the pollutant has been eliminated.

Example At noon, Smith's house contains 1 gram of helium. No more helium is added. A small exchanger is started up at noon and provides, by the end of a 3-hour period, a replacance of 90%. What is the concentration of helium in the house at that time? Answer: (1 - 0.90) = 10% of the initial amount. In short, 0.1 gram remains.

A more useful formula is this: $C = C_i (1 - \text{replacivity})^P$ where P is the ratio of time interval to one hour.

__Example__ At noon, Smith's house contains 16 grams of xenon. No more is added. An air-to-air heat-exchanger that has a replacivity of 50% is started up and run for two hours. What is the concentration of xenon at the end of that period? Answer: $16 \text{ g } (1 - 0.50)^2 = 16 \text{ g } (0.25) = 4$ grams.

CASE 2: CONTINUOUS INTRODUCTION OF POLLUTANT

Here I assume that more of the specified pollutant is introduced continually and that the rate of introduction is constant. I call this rate R_{ap}, where __R__ stands for __rate__ and __a__ and __p__ stand for __added__ and __pollutant__. Rate may be expressed in grams per hour, or other unit per hour. I assume that the exchanger is running steadily, and I assume that these processes (addition of pollutant and operation of the exchanger) have been going on for such a long time that the concentration of the pollutant has become steady, i.e., has reached an equilibrium value, which I call C_{eq}. (Of course, if there were no change of air, the concentration would continue to increase indefinitely.)

Clearly, the greater the rate of fresh air input by the exchanger, the lower the equilibrium concentration C_{eq}. How, exactly, does C_{eq} depend on the air input rate? The answer is easily found, easily stated:

$$C_{eq} = \left(\frac{1}{\frac{\text{rate of air input, in house}}{\text{vol. per hr}}} \right) (R_{ap}) \qquad = \qquad (1/R_i)(R_{ap}) \qquad = R_{ap}/R_i \; .$$

The validity of this formula is apparent from consideration of the case in which 1 gram of pollutant is added each hour and one house-volume of fresh air is injected each hour (and, of course, one house-volume of air is __ejected__ each hour). Steady and ideal mixing of indoor air is assumed. Then if the concentration of pollutant is to remain constant (i.e., if there is to be equilibrium between input and outgo of pollutant), the quantity of air ejected each hour must contain just as much pollutant as is added to the indoor air per hour, i.e., must contain 1 gram of pollutant. But the ejected air will carry this amount of pollutant only if the indoor concentration is one gram per house volume.

__Example__ If R_{ap} is 3 grams/hr. and R_i is 2 house-volumes per hour, then
$C_{eq} = R_{ap}/R_i = 3/2 = 1.5$ grams per house volume.

The following table shows representative values of equilibrium concentration for various rates of fresh air input. The equilibrium concentration is expressed in terms of R_{ap}, the hourly rate of addition of pollutant. This rate may be stated in terms of gram/hr, liter/hr, microcuries/hr, or any other "per hour" unit.

R_i (house volumes per hr)	C_{eq}
1/8	$8 \; R_{ap}$
1/4	$4 \; R_{ap}$
1/2	$2 \; R_{ap}$
1	$1 \; R_{ap}$
2	$0.5 \; R_{ap}$
4	$0.25 \; R_{ap}$
8	$0.13 \; R_{ap}$
16	$0.06 \; R_{ap}$

EFFECT OF NATURAL INFILTRATION

Obviously, if an exchanger-equipped house has a high rate of natural infiltration, the formulas, tables, etc., presented above do not apply. Because of the natural infiltration, the concentrations of pollutants will be lower than if fresh air were introduced by the exchanger only. How much lower? In rare cases one may be able to answer this question in rigorous manner. Usually one can do little more than guess.

Chapter 9

ENERGY, HEAT, AND HEAT-FLOW: SOME BRIEF NOTES

Introduction

Energy

The great embarrassment concerning energy

Heat

Temperature

Enthalpy

Heat flow

INTRODUCTION

This is not a book on energy or heat. Nevertheless some facts about these basic concepts, and about enthalpy, are worth including. Enthalpy is unfamiliar to many people, yet it is easy to understand; it plays a dominant role in discussions of exchangers that transfer both heat and moisture.

ENERGY

The very foundation of physics and engineering is energy. It is all around us. No action can be initiated without it. It is the common currency of action and change. Equations and formulas involving energy can be simple because of this simplifying fact, or law: the overall amount of energy in any system (any closed system) is constant. It cannot be increased. Cannot be decreased. It remains the same forever. This gives physicists a big headstart at writing formulas involving energy.

What is energy, exactly? It is the capability to do something; for example, to lift a heavy weight, accelerate an automobile, produce electricity, warm some air, operate a blower, or evaporate some water. A kind of ultimate "acid test" for energy is: can it impart heat to some extremely cold material? Only energy fills this bill. (Speed, momentum, force, etc. cannot impart heat to a cold material. Only energy can.) If the acid test is to give strictly accurate results, the extremely cold material must be at a temperature almost as low as absolute zero; but in most instances one can get by with the use of very cold water.

Kinds Of Energy

There are a great many kinds of energy. From time to time, additional kinds have been discovered, for example, nuclear energy, mass energy. Some of the commonest kinds are heat, energy of motion (kinetic energy), gravitational potential energy, chemical energy, electrical energy, radiant energy (associated with visible radiation, ultraviolet, x-rays, gamma-rays, infrared, radio waves).

<u>Units Of Energy</u>

The commonest units are:

> British thermal unit (Btu), defined as the amount of energy needed to raise the temperature of one pound of water one degree F. Originally applied just to heat, but later applied to all kinds of energy.
> Joule (J), defined as one watt-second. Applicable to all kinds of energy. In most countries of the world it is the preferred unit.
> 1 Btu = 1055 J. 1 J = 9.478 x 10^{-4} Btu.
> 1 Btu = 2.931 x 10^{-4} kWh. 1 kWh = 3412 Btu.

THE GREAT EMBARRASSMENT CONCERNING ENERGY

The concept of energy has this Achilles' heel: to compute (or merely to define!) the total energy of a system -- including <u>every</u> kind of energy it may contain -- is virtually impossible.

In calculating the energy of a system, an engineer may arrive at very different answers depending on how many kinds of energy he chooses to include. If he is interested in heating a house deep in the woods of Vermont, he may include only (a) sensible heat, (b) the chemical energy in firewood, and (c) radiant energy from the sun.

Also, he may get very different answers depending on what assumptions he makes as to the surroundings, that is, assumptions as to what the energy is <u>relative to.</u> Does a small stone beneath my foot have zero potential energy? Does your answer change if I tell you that I am standing high up on a mountain?

From decade to decade new forms of energy are discovered. An engineer back in 1900, on inspecting a hunk of uranium, would have declared it to have almost no energy. Today, every engineer would give a very different answer. Thus our appraisals of the energy of a system may change from decade to decade as our understanding of the physical universe grows.

<u>Limitations In Interest</u>

Seldom are persons interested in the <u>total</u> amount of energy in a system; they are usually interested in one kind of energy only -- for example heat energy, or electrical energy. And, even with respect to one kind of energy, they are seldom interested in the <u>total amount</u>; they are usually interested in energy changes, energy increments; they want to know "How much energy does this process add, or take away? How much energy can be recovered?"

<u>Challenge</u>: Leaf through a half-dozen books on physics and see whether any of them ever discuss <u>true total amount</u> of energy. I predict that none of them do. Or ask any physics professor: "What is the total amount of energy in one pound of 70F water in my kitchen?" I predict that you will get a lecture, but no number.

<u>The Consequence</u>

The consequence of these embarrassments concerning energy is that engineers usually confine their attention to <u>limited</u> classes of energy and to <u>increments</u> within such classes. For example, they may deal just with heat, or, more exactly, additions and subtractions of heat. They may deal also with so-called latent heat, or internal energy, or enthalpy (defined in a later paragraph).

HEAT

This is one of the commonest forms of energy. It is the energy associated with the random motions of atoms and molecules. The more rapidly they move about and vibrate and rotate, the more heat they contain. Note that, because heat is a combination of myriad random motions, heat cannot be manipulated in a simple and concise way; thus it differs importantly from light, electricity, etc., which we can control and transform with great elegance. Always heat is a grand mixture of kinds and quantities of motion; it can be dealt with only in an overall, statistical way; it is, in a sense, a vulgar, intractable kind of energy.

Heat is called an <u>extensive</u> quantity: the amount of heat in a system depends directly on size of the system. Double the size, and you double the amount of heat. The amount of heat in the oceans is enormous, even though they are only luke warm at best! The amount of heat in a cup of boiling water is trifling, even though the temperature of the water is very high.

Comment On The Error Of Calling Infrared Radiation Heat

Electromagnetic radiation (light, infrared, radio waves, etc.) is not heat. Electromagnetic radiation can be manipulated in a concise way: it can be reflected, focused, polarized, etc. Heat cannot. Many authors who write about infrared radiation commit the crime of calling it a form of heat. It is not heat. No electromagnetic radiation is heat. Any kind of such radiation, when absorbed, can <u>produce</u> heat, i.e., can be converted to heat. In this respect, infrared radiation is no different from light, radio waves, etc.: they too, on being absorbed, are entirely converted to heat.

Symbol For Heat

The amount of heat in a system (more exactly, the amount above some arbitrary reference level) is given the symbol Q, for <u>quantity</u>.

TEMPERATURE

Temperature is a measure of the violence of the random motions of atoms and molecules. It is an intensive parameter, not an extensive one. That is, it is not influenced by how much matter you have, but only by the violence of the random motions. The commonest units of temperature are: degree Fahrenheit (F or $^{\circ}$F) and degree Celsius (C or $^{\circ}$C). 32°F and 212°F correspond to 0°C and 100°C respectively. To convert from Fahrenheit to Celsius, subtract 32 and multiply by 5/9.

Absolute Scales

The Rankine scale is like the Fahrenheit scale but starts at absolute zero. Thus the Fahrenheit values 0, 32, 70, 100, correspond to 459.67, 491.67, 529.67, and 559.67 Rankine. To convert to Rankine, add 459.67. The Kelvin scale is like the Celsius scale but starts at absolute zero. Thus 0, 20, and 100 $^{\circ}$C correspond to 273.15, 293.15, and 373.15 K.

ENTHALPY

Enthalpy is to heat as apple-pie-a-la-mode is to ordinary apple pie. It is a combination: an especially happy combination. It is the combination of <u>heat energy</u> and "<u>pressure-volume</u>" energy. Called H, it is expressed as:

$$H = U + pV$$

where U is the heat (also called -- not very appropriately -- internal energy), p is the pressure of the system, and V is the volume of the system.

One assumes that the system consists just of a single small chamber containing just gas -- so that one value of pressure characterizes the entire system.

Enthalpy is a key, or <u>the</u> key, parameter in many calculations involving energy transfer, energy recovery, energy saving, specifically those calculations that involve the combination of heat and pressure-volume energy. If a quantity of air is warmed up, the heat is increased. If a small amount of water evaporates into this air, the pressure-volume energy is increased. If both processes occur at once, both kinds of energy are increased -- which may be summarized by saying that the enthalpy increases. And it is this overall quantity -- enthalpy -- that really counts. Why? Because, when the two kinds of energy changes occur, the law of conversation of energy <u>does not apply to either alone</u>: it applies only to the combination, i.e., the enthalpy. The "bottom line" is enthalpy. What the designer is trying to conserve, ordinarily, is enthalpy.

Warning: This is not always the goal. If, in winter, a house contains much air that is warm, stale, and too humid, the occupant's goal, in using a heat-exchanger, is to bring in fresh air and expel (not save!) moisture. He desires to expel moisture even if this entails a loss of enthalpy.

HEAT FLOW

The total amount of heat that has flowed from one body to another in a given time interval is called Q, and the amount that flows each second is called q. Btu is a common unit of amount of heat.

Heat may flow by any and all of these three mechanisms: (1) conduction, in which the matter stays roughly fixed but the energy (kinetic energy of atoms and molecules) moves along, (2) convection, involving gross travel of material, e.g., water or air, and (3) radiation.

Everything is emitting electromagnetic radiation at all times (everything that is hotter than absolute zero) and in all directions. Likewise everything is receiving and absorbing such radiation at all times. Emission tends to make an object cooler, and absorption tends to make it hotter. What counts, usually, is the difference between amount emitted and amount absorbed. Often this difference is very small, in which case heat-flow by radiation can be disregarded.

Chapter 10

AIR-TO-AIR HEAT EXCHANGERS OF FIXED TYPE: DESIGN PRINCIPLES

Introduction

Principle of operation

Exchanger proper

Three choices of flow-direction relationship

Steady-state vs. reversing-flow exchangers

Laminar flow vs. turbulent flow

Flowrate required

Choice between blower and fan

Manifolds

Pressures and flowrates

Insulation

Orientation

Operating procedure

Temperature distribution along exchanger proper

Exchangers in which transfer of water occurs

Could exchanger moisture-recovery be controllable?

Exchangers employing heat-pipes

Use of preheat and postheat

Means of controlling frosting

INTRODUCTION

Here the general principles of design of air-to-air heat-exchangers of fixed type (non-rotary) are set
forth.
      In most exchangers the heat-exchanger sheets are impermeable to water (gaseous or liquid water)
with the consequence that no transfer of water from one airstream to the other can occur. These
exchangers are discussed first.

PRINCIPLE OF OPERATION

In an air-to-air heat-exchanger of typical design there are two streams of air and these are separated
by a set of thin sheets of metal or plastic through which heat can flow easily but no air can flow, and
no water, whether in liquid or gaseous form, can flow.
      Each stream is driven by an electrically powered blower (or fan). Usually two blowers are used.
Some exchangers employ one blower and one fan.

The intervening thin sheets may be flat, may be curved to form tubes, may be accordion folded, or may have other shapes. They may be called sheets, plates, membranes, septums, tubes, pipes, or ducts. The set, aggregate, or assemblage of such devices is called the exchanger proper, or core.

Because a stream of warm outgoing air "bathes" one side of each sheet and cold incoming air bathes the other side, one side is hotter than the other and accordingly heat flows through the sheet. By the time the outgoing air reaches the outdoors it has given up so much heat that is is nearly as cold as the outdoor air; correspondingly the incoming air, as it enters the room, is nearly as warm as the indoor air. Overall, about 60 to 80% of the heat that would have been lost is recovered. (See Chap. 14 for heat-flow laws and formulas.)

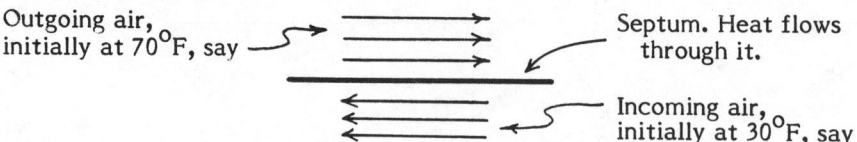

The blower or fan in a given airstream may be situated upstream or downstream from the heat-transfer sheets. That is, it may be pushing or pulling air through the exchanger proper.

Filters may be provided in either or both of the airstreams. Usually each filter is located up-stream from the exchanger proper; thus it prevents dust from entering the exchanger proper and perhaps lodging there and restricting the slender passages for airflow and slightly reducing the rate of heat-flow through the thin sheets. The filters may have to be cleaned periodically.

The outdoor discharge location of the outgoing air is kept well separated from the outdoor air-intake location. Accordingly no appreciable amount of discharged air is reintroduced to the house.

The air to be discharged from the house may be drawn from one room (e.g., kitchen, bathroom, or other room where pollutant concentration may be especially high), or from several rooms. Like-wise the incoming fresh air may be delivered to one room (e.g., living room or other room where the house occupants spend much time) or to several rooms.

EXCHANGER PROPER

Many air-to-air heat-exchangers employ flat sheets. Some employ concentric cylindrical shells. Some employ tubes or heat-pipes. One well known type employs accordion-folded sheets. Here I make some comments on sheet-type and tube-type exchangers.

A typical sheet-type exchanger includes a large number of sheets arranged in a stack, with planar airspaces between sheets. Incoming air flows in Spaces 1, 3, 5, etc., and outgoing air flows in Spaces 2, 4, 6 etc. Thus the total area of heat-exchange surface may be very large while the length and breadth of the assembly are small.

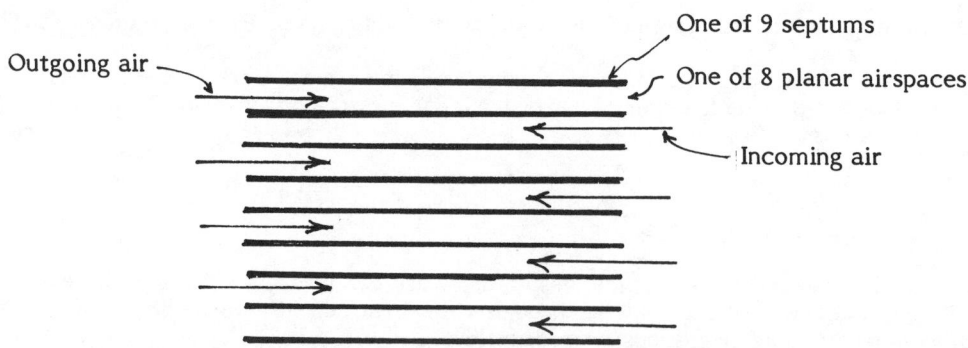

Cross-section of a portion of air-to-air heat-exchanger of multi-flat-septum type. For simplicity, the two sets of headers, or manifolds, are not shown.

A tube-type exchanger may include many tubes running parallel to one another, with spaces between. Outgoing air (say) flows in the tubes and the incoming air flows (in antiparallel direction or cross direction) in the spaces between them. Again a large area of heat-transfer is provided in a compact region.

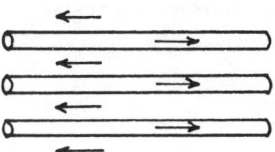

To save space and to improve heat-transfer, the designer usually chooses very small spacings and very small diameters, as suggested by the following drawings.

Flat-septum type   Tube type

THREE CHOICES OF FLOW-DIRECTION RELATIONSHIP

An interesting topic is the choice of relationship of the directions of the two airstreams. There are three distinct choices, each with its own advantages and disadvantages.

Concurrent-flow   Here the two streams have the same direction.

Crossflow   Here the two streams are at right angles to one another.

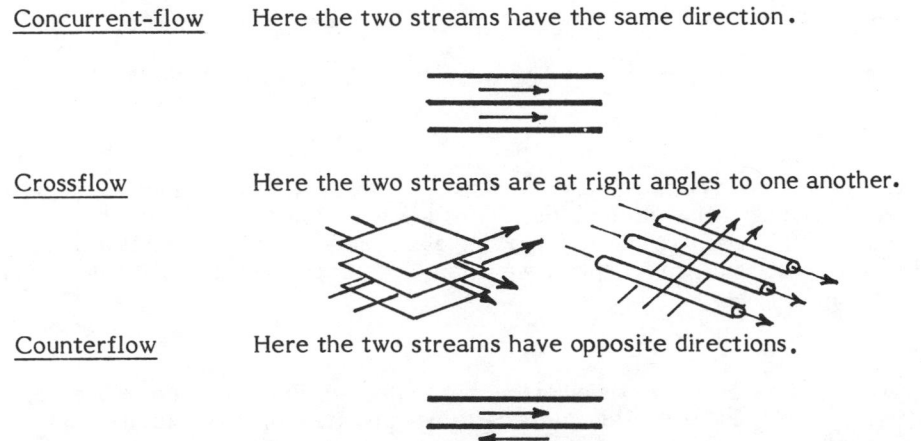

Counterflow   Here the two streams have opposite directions.

Counterflow exchangers can have very high efficiencies even when of modest overall dimensions-- and, in principle, can have near-100% efficiency if the heat-exchanger area is extremely large. Cross-flow exchangers also can have very high efficiency, but this is attainable only if the heat-exchanger area is somewhat larger than that of a counterflow exchanger of equal efficiency. Concurrent flow exchangers have very low efficiency (well below 50%, usually) and are not discussed in this book.

46

STEADY STATE VS. REVERSING-FLOW EXCHANGERS

Steady-State Exchanger

In most exchangers, operation is steady. The two airflows are constant. No mechanical part (other than the blowers) moves. Heat enters each septum via one face thereof and emerges from the other face; all of the heat that is transferred has traveled from one set of airspaces to the other -- has traveled through a septum.

Reversing-Flow Exchanger

It is possible to design an exchanger that has only one set of spaces: the two airflows follow one another sequentially. They take turns. The direction of airflow reverses periodically, say every minute; the flow is outward for one minute, then inward for one minute.

In such an exchanger there is no heat-flow through anything. Heat flows (a short distance) into various solid objects (airspace walls, etc.), then flows back out of these objects. It flows out via the same surfaces that, earlier, it flowed in by.

Because the heat does not flow all the way through anything, the various airspace walls, etc. do not need to be very thin and do not need to have very high thermal conductivity.

Furthermore, heat-transfer no longer takes place just at walls: it can take place also at such strips, rods, wires, etc. as may have been installed wholly within the air passages. Thus the designer can provide a very large heat-transfer surface by simple means and at low cost. Only the envelope proper needs to be airtight. The strips rods, or wires may be of cheap material crudely fabricated and crudely mounted.

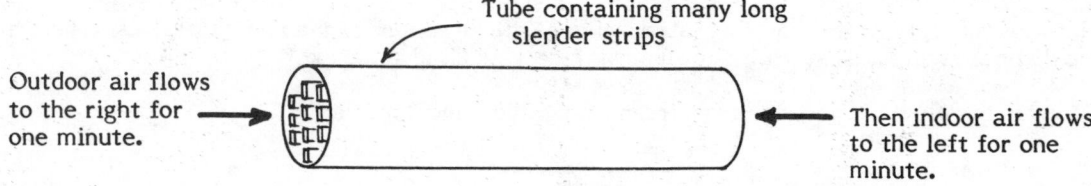

Such assemblies can be used in pairs employed in opposite phase: at any one moment there is air outflow in one assembly and air inflow in the other. Thus there is no net build-up or build-down of air pressure in the house. (It is also possible to use a single assembly. If the flowrates are low, if the flow is reversed every few seconds, and if the house is large, the rapid succession of small changes in pressure in the house may be acceptable.)

Rotary Exchangers

One special type of reversing-flow exchanger is a device that employs a slowly rotating wheel, or rotor, that lies athwart two parallel side-by-side ducts carrying oppositely directed airstreams. Such exchangers are so important that a separate chapter is devoted to them; see Chapter 12.

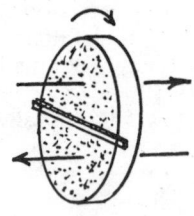

LAMINAR FLOW VS. TURBULENT FLOW

Introduction

It is well known that when air is travelling at low speed along a small-diameter, smooth-walled tube, the flow is likely to be laminar; that is, each molecule of the air tends to move in approximately a straight line parallel to the tube.

Laminar flow

But if the linear flow-speed is high and the tube is fat and the walls are rough, the flow is likely to be turbulent, i.e., likely to include a great number of whirls, eddies, etc.

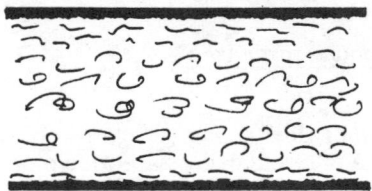

Turbulent flow

The same applies to air that is flowing in the planar spaces between parallel plates, or in spaces of other shape.

Why Is The Choice Of Flow-Type A Difficult One?

It is difficult because:

The amount of power needed to drive air through a given set of passages is less if the flow is laminar. The air travels much more easily. Smaller, lower-power blowers may be used. Electrical power consumption is less. The airflows are quieter.

The extent of heat-exchange is greater if the flow is turbulent -- provided that the heat-transfer area and flowrates are the same. When the flow is turbulent, near-stagnant ("insulating") airfilms close to the heat-exchange surfaces are much thinner and play only small roles. Heat exchange is maximized.

Usually the designer wishes the exchanger to be compact: to have a small volume. Also, he wants the electrical power consumption to be small. Accordingly his usual choice is laminar flow.

How does one keep the flow laminar?
By (1) using slender passages, (2) keeping the linear speed of airflow low, and (3) keeping the passage walls smooth and free of obstructions.

Exactly how slender should the passages be, and exactly how low should the linear speed of airflow be? The answers are easily obtained by computing a kind of "index number of tendency toward turbulence", called Reynolds number. This is defined in a following paragraph.

How does one detect turbulence?
To detect turbulence visually, one can make the exchanger of transparent material and introduce some smoke or white dust into the input air. One can then see the flow. Proper illumination, for example darkfield illumination such as is used routinely by microscopists, is helpful. If the flow characteristics can be seen, they can also be photographed.

48

Change of flow type may be detected by monitoring sudden changes in flowrate or in pneumatic resistance or in blower electric power level. A change from laminar to turbulent flow may be evidenced by a sudden decrease in flowrate, sudden increase in resistance, or sudden increase in electric power used by the blower.

Definition of Reynolds Number

The subject of flow of gases or liquids is dominated by Reynolds number, which is a kind of index, or measure, of the tendency of a flow to be turbulent. The number is a kind of ratio: ratio of (1) product of parameters which, if large, tend to make the flow turbulent, to (2) the single parameter which, if large, tends to make the flow laminar. In this ratio, the dimensions cancel out: Reynolds number is dimensionless. Thus it applies equally to English units or metric units or other units.

The symbol for Reynolds number is Re.

The quantity is defined as follows -- with respect to the flow of air:

$$Re = \frac{\rho u_m \delta}{\mu}$$

where:

ρ is the density of air, in $\frac{(pound\ force)(sec^2)}{ft^4}$.

u_m is the average linear speed of flow -- an average taken over the cross section of the air passage (a circular-cross-section pipe, for example).

d is the diameter of the air passage (pipe). Unit used: ft.

μ is the viscosity (absolute viscosity, or dynamic viscosity) of the air.

Unit used: $\frac{(pounds\ force)(sec)}{ft^2}$.

As indicated in Chapter 2, the density of standard-pressure 70F air is 0.077 lb. mass/ft^3, or 0.00239 (lb. force)(sec^2)/ft^4, and the viscosity of such air is 3.8×10^{-7} (lb. force) (sec)/ft^2.

Example Suppose 70°F standard-pressure air is flowing at 10 ft/sec. in a circular-cross-section duct that is 4 in., i.e., 4/12 = 0.333 ft., in diameter. Then the Reynolds number is

$$Re = \frac{(0.00239\ lb.\ force.sec^2/ft^4)(10\ ft/sec)(0.333\ ft)}{3.8 \times 10^{-7}\ lb\ force,\ sec/ft^2} = 21,000 \text{ (dimensionless)}$$

Discriminating Value Of Reynolds Number

Countless experiments have shown that if Re exceeds 2000, the flow is likely to be turbulent -- very likely if Re exceeds 3000. If it is below 2000, the flow is likely to be laminar.

If the number is in the neighborhood of 2000 or slightly greater, the flow type cannot be predicted with confidence. The exact extent or type of roughness of the tube walls and certain other circumstances may govern. For a detailed account of roughness and its influence, see L.S. Marks "Mechanical Engineers' Handbook", 5th ed., p. 249. The roughness of tubes of glass, drawn brass, drawn copper is negligible -- when the tubes are new and clean. The roughness of galvanized steel pipes, and of many other types of passages, may be significant -- too great to be neglected -- if the Reynolds number is in the range 2000 to 3000.

Example of laminar airflow in circular-cross-section duct Consider air that is flowing at 10 ft/sec. in a duct that is 0.3 in. in diameter. The Reynolds number is about 1500 -- well below the discriminating value of 2000 to 3000. Thus the flow is laminar.

Application To An Airspaces Between Parallel Plates

Suppose that, in an exchanger containing many parallel plates, a typical air passage has a (rectangular) cross section 12 in. x ¼ in. and air if flowing along the passage at 5 ft/sec. Is the airflow laminar?

     To find the answer, one again calculates the Reynolds number and compares it with the discriminating value. But, in making the calculation, what number is to be used in place of <u>diameter</u>? As explained in various standard textbooks, for example Coren's "Chemical Engineering" (C-720), the procedure is to find the <u>equivalent diameter</u> which, on the basis of long experimentation, has been found to be about (4/S)(cross sectional area), where S is the sum of the lengths of the four lines comprising the duct cross section -- in this example, 12 + 12 + ¼ + ¼ = 24½ in. The cross-sectional area, in this example, is 12 x ¼ = 3 sq. in. Thus the equivalent diameter is $[4/(24½ \text{ in.})](3 \text{ in.}^2)$ or 0.041 ft. Noting that the speed of airflow is 5 ft./sec., one finds the Reynolds number to be about 1300. Thus the flow is laminar.

Why Is The Calculation, Or Prediction, Of Flow-Type So Difficult?

Because:

    The scientific foundation is complicated. It involves many parameters. Some of these are unfamiliar to most architects and builders.

    To some extent the subject is an art, not a science. This applies especially to an exchanger that operates close to a "transition zone". The fact is, there is a fairly wide transition zone -- a fairly wide range of parameters where the flow type is not accurately predictable and the behavior is delicate, unstable, fickle.

    One of the parameters that affects the type of flow -- surface roughness -- is hard to express in businesslike way. There are so many kinds (and extents) of roughness!

    Because air is invisible, and because exchangers usually consist of opaque sheets and tubes, an experimenter has a hard time finding whether, in each part of the exchanger, the flow is turbulent or laminar. In short, diagnosis is difficult.

FLOW-RATE REQUIRED

Flow-rate requirements are discussed in detail in Chapter 15. Typically, an average-size house that has been fairly tightly built may need an exchanger that will provide 50 to 150 ft^3 of fresh air per minute. Of course, non-tight houses may need no exchanger at all, and a house that is very tight and contains many sources of pollutants (people smoking, people taking long showers, cabbage being cooked, large areas of materials that contain formaldehyde, gas stove defective) may need a higher rate of fresh-air input.

    Ideally, the rate of fresh-air input should be varied according to the need, as has been pointed out by Woods et al (W-450). This topic also is discussed in Chapter 15.

Should The Two Flowrates Be Equal?

Usually they should be. Most exchangers are designed to produce equal rates of fresh-air input and stale-air output.

Difficulty In Keeping The Two Flowrates Equal

It is explained in Reference F-70 that keeping the two flowrates equal is often difficult. If there is slight deformation of the walls in one airstream but not the other, or if the flow-paths in one stream are more circuitous than those in the other airstream (in the header, or manifold, regions, say), or if one circuit has greater accumulation of dust or frost than the other has, the two flowrates may differ significantly. Differences in resistance along the two sets of ducts (serving the two air-streams) likewise may produce imbalance of flow. One company insures keeping the two blower speeds identical by driving the blowers with a single motor.

CHOICE BETWEEN BLOWER AND FAN

Blowers are preferred when the pressure head is large. Fans, sometimes called axial blowers, are preferred when the pressure head is small. Most air-to-air heat-exchangers employ blowers.

Blower   Fan

Blowers are of centrifugal (squirrel-cage) type. The vanes that drive the air may be radial or may have forward (positive) or backward (negative) slant, positive slant being preferred it the pressure head is relatively small.

Forward (positive) slant
of rotor of blower

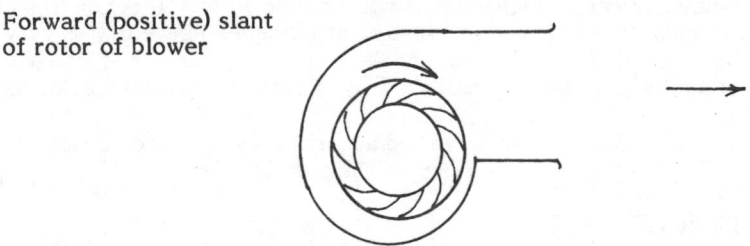

MANIFOLDS

Specially shaped manifolds, or headers, are employed to introduce the airstreams to the pertinent passages within the exchanger proper and to receive the emerging air. The point is, of course, that the two streams must be kept separate at all times. If they become mixed, much of the air heading toward outdoors may soon find itself en route back to the rooms.

Designing the manifolds is sometimes difficult (and making clear drawings of this is also difficult).

If many airstream-direction changes are caused by the manifolds, the resistance to airflow is increased and higher power blowers may be required.

Later chapters show the design of many kinds of manifolds.

PRESSURES AND FLOWRATES

See Chapter 13.

INSULATION

Most heat-exchanger outer surfaces, or housings, are insulated -- to minimize flow of heat to or from the general surroundings. Of course, such insulation is of small importance if the exchanger is very compact, i.e., has a small external surface area despite having a large total heat-transfer area. Ordinarily, however, the use of insulation is cost-effective. Insulation applied to the exchanger housing may be applied to the inner or outer face of the housing. Ducts are usually insulated externally. The commonest insulating material used is fiberglass.

ORIENTATION

Most exchangers can have a variety of orientations, subject only to the arrangement of drains for any liquid water that is produced. As far as airflow and heat-exchange are concerned, the airflow directions in the exchanger proper may be east and west, north and south, up and down, or at any angle from such direction.

OPERATING PROCEDURE

The simplest procedure is to let the exchanger run continuously throughout the midwinter period -- period when windows are kept closed. The cost of the electrical power for the blowers may be near-negligible, and the amount of heat lost (if, say, the exchanger has an efficiency of 70% instead of 100%) may be acceptable.

Next simplest is to have the house occupants control the exchanger manually: turn it on when they deem appropriate and turn it off when they deem appropriate. Their actions are governed by their sensations as to whether moisture, smells, etc., are excessive and on what they have heard or read as to the threats posed by formaldehyde, radon, etc. But if the occupants are careless or forgetful the overall performance may be poor.

Automatic controls can be used, and, in principle, can regulate pollutant concentrations more reliably and more accurately. Sensors that respond to different levels of humidity already exist and are relatively simple. But to provide a set of sensors that respond to many kinds of pollutant (smells, formaldehyde, radon, humidity, etc.) would be very expensive. Perhaps inexpensive sets will be developed in the next few years.

The need for better sensors and controllers has been stressed by Woods, Maldonado, and Reynolds (W-450). See also Chapters 15 and 16.

TEMPERATURE DISTRIBUTION ALONG EXCHANGER PROPER

In each airstream of a fixed-type exchanger there is a temperature distribution along the exchanger, i.e., along the air in any given slender passage. The distribution depends on many design parameters and also on the flowrates of the two airstreams.

If the two rates are equal and are extremely low, and if no condensation occurs, the temperature of the incoming air at successive locations along the exchanger (starting, say, from the cold end, here assumed to be the left end) is successively higher and the temperature of this air as it enters the room is almost as high as room temperature. The temperature of the outgoing air has a nearly identical distribution (again I assume that distance is measured from the cold end -- left end). In other words, at any given location along the exchanger, the outgoing air has slightly higher -- but only slightly higher -- temperature than the incoming air (at this location). The two distribution curves (see below) are nearly straight and they have (at any one location along the exchanger) identical slope, as must be the case inasmuch as the heat that is lost by one airstream is gained by the other. Obviously, the efficiency of sensible heat recovery, in the case under discussion, is very high. In drawing the graph I have assumed that the indoor and outdoor temperatures are 70°F and 30°F, the exchanger is 20 in. long, and the heat-exchanger efficiency is about 90%.

Case 1
Very low flowrate,
high efficiency

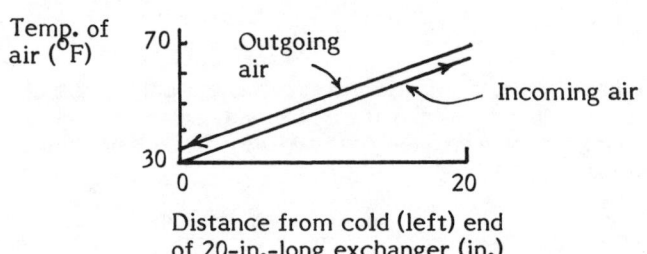

Distance from cold (left) end
of 20-in.-long exchanger (in.)

Note: Each of the two curves may be slightly steeper at the left and right ends if the airflows, although mainly laminar, are somewhat turbulent where the outgoing air enters the exchanger (warm end) and the incoming air enters the exchanger (cold end). The turbulence partly eliminates the stagnant boundary airfilms and thus increases (very locally) the rate of heat-transfer. See following graph.

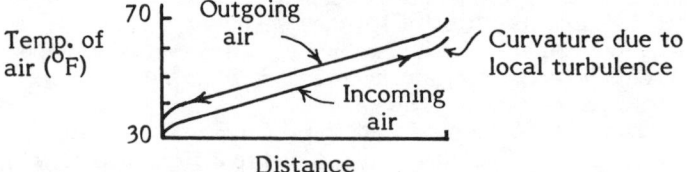

If the two rates of airflow are extremely high, the two distribution curves are again approximately straight lines and the slopes are identical, but the two curves are far apart: the incoming air never gets very warm and the outgoing air never gets very cold. The efficiency of sensible heat recovery is low -- about 25% in the case depicted below.

Case 2
Very high flowrate,
low efficiency

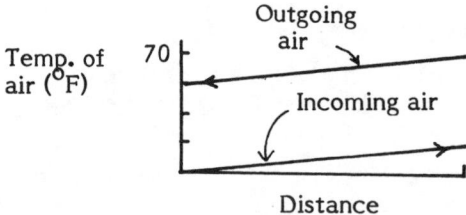

In practice the rates of airflow are chosen so as to provide a good compromise between the desire for high efficiency and the desire for a rapid rate of fresh air input. Achieving a good compromise is facilitated if the blowers are of variable-speed type, i.e., if the flowrates can be altered to accommodate varying circumstances.

Note Concerning Symmetry

It is a great help, when trying to infer the shapes of the temperature distribution curves along an exchanger that has been designed to recover sensible heat only, to know that, under circumstances such that no condensation occurs, the combination of curves for incoming air and outgoing air must be right-vs-left-and-up-vs-down symmetric. That is, if the pair of curves is rotated about a horizontal line and then rotated about a vertical line, the result is the same as at the outset. All this is a consequence of the fact that the main heat-transfer processes are linear or nearly linear and the algebra of temperature differences and heatflow quantities is indifferent to the <u>signs</u> of the differences.

If some condensation occurs in the outgoing air, the situation is different. Each of the two temperature distribution curves consists of <u>two</u> straight segments and these have different slopes. The ("incoming" and "outgoing") segments at the right (those pertaining to the warm end of the exchanger) are parallel because no condensation occurs here and the temperature gain of the incoming air must be equal to the temperature loss of the outgoing air. But the <u>left</u> segments are not parallel; the temperature gain of the incoming air far exceeds the temperature loss of the outgoing air inasmuch as this latter air loses energy by <u>two</u> mechanisms one of which (condensation) does not necessarily involve any decrease in temperature. (In drawing the graph I have assumed typical flowrates.)

Case 3
Condensation occurs

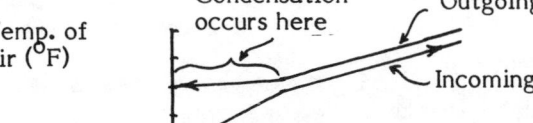

If one prepares a graph of <u>enthalpy</u> vs. distance, instead of temperature vs. distance, one finds the two curves to be parallel even in the left portion of the graph. Why? Because the enthalpy loss of the outgoing air equals the enthalpy gain of the incoming air irrespective of whether condensation occurs. (Enthalpy and condensation are discussed in Chapter 11.)

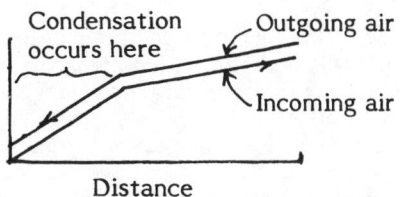

EXCHANGERS IN WHICH TRANSFER OF WATER OCCURS

In some exchangers the heat-transfer sheets are water-permeable. Thus gaseous water (and liquid water also, I assume) can transfer from the stream of outgoing air to the stream of incoming air -- which is desirable if room air threatens to be too dry and the outdoor air is very dry. Such an exchanger recovers not only sensible heat but also latent heat. One type of Mitsubishi exchanger has water-permeable heat-transfer sheets. See Chapter 20.

Some rotary exchangers (exchangers employing a massive rotating wheel) include, in the massive rotor, a desiccant. This effectively transfers water from one airstream to the other; see Chapter 12. Several types of exchangers marketed by Berner International Corp. have this capability; see Chapter 19.

Even if the rotor contains no desiccant, transfer of water from one airstream to the other will occur if there is some condensation of water in the outgoing airstream. The resulting water may be picked up by the incoming air.

COULD EXCHANGER MOISTURE-RECOVERY BE CONTROLLABLE?

In winter, a house occupant may sometimes want the exchanger to recover moisture (to save latent heat and to keep the room air from becoming too dry) but at some other times he may want it not to recover moisture, because room air tends to be too humid.

How nice it would be if, at the touch of a control button, he could switch from "moisture recovery" to "moisture non-recovery"! Offhand, I can think of no easy way of making such change. However, this possibility deserves some thought: the manufacturer could supply -- not one -- but two exchanger cores: one employing permeable sheets and the other employing impermeable sheets (or one employing desiccant and the other desiccant-free). The house-occupant could remove one core and substitute the other. (There is at least one make of exchanger (Mitsubishi Lossnay VL-1500) designed so that the core can be removed in about one minute, without tools.)

Conceivably an exchanger could be designed so that merely turning a knob would swing one core out of the airstreams and swing the other into the airstreams. Perhaps the knob could be turned by an electrical device controlled by a dehumidistat.

I have recently hit upon some schemes that might be even simpler. They are described in a report dated 12/12/81. See Bibliography item S-235-mm.

EXCHANGERS EMPLOYING HEAT-PIPES

See Chapter 21, which describes Q-Dot exchangers. They include heat-pipes. See also pages 3-17 of Reay's "Heat Recovery Systems: a Directory of Equipment and Techniques", Bibl. item R-25.

USE OF PREHEAT AND POSTHEAT

If one adds a little heat to the incoming fresh air before this air enters the exchanger proper, one enjoys two benefits: (1) condensation and frost formation may be avoided despite very low outdoor temperature, and (2) the incoming air is somewhat warmer as it enters the room and thus avoids creating discomfort.

Several manufacturers of exchangers recommend preheating when the outdoor temperature is low.

Of course such preheating imposes a penalty: the incoming air picks up less heat from the exchanger proper. That is, the efficiency of heat transfer is reduced.

If one postheats the incoming air (adds heat after it has passed through the exchanger proper), the above-mentioned penalty is avoided. Also, the virtue of avoiding discomfort to persons in the room is retained. However, no benefit with respect to avoidance of condensation and frosting is realized. Accordingly postheating is seldom used -- except when the furnace is of hot-air type and it is convenient to distribute the incoming fresh air via the furnace hot-air system; this air then picks up heat whenever the furnace is running.

MEANS OF CONTROLLING FROSTING

See following chapter.

Chapter 11

WATER VAPOR, HUMIDITY, CONDENSATION, AND WATER TRANSFER

INTRODUCTION

Here I deal with liquid and gaseous water, humidity, condensation, water transfer, and frosting. These subjects are important with respect to exchangers that do not exchange latent heat and even more important with respect to those that do exchange latent heat and do exchange water.

Much attention is given to the Multi-Purpose Humidity Graph, or Psychrometric Chart. Anyone who is really familiar with this graph can solve countless problems involving sensible heat, latent heat, etc., almost "in his head." Problems that might otherwise seem inscrutible, overwhelming, depressing may become easy, obvious, fun. To paraphrase a famous witticism: Using this graph we can see the unseeable, hear the unhearable, and unscrew the inscrutible!

THE MULTI-PURPOSE HUMIDITY GRAPH (PSYCHROMETRIC CHART)

This graph, shown in all its glory on a following page, is so complicated looking that, initially, I had much trouble in understanding it. Perhaps many other persons also have had trouble. To make the graph easy to understand, I introduce it, or develop it, in several small steps.

Step 1 Construct a rectangular grid the horizontal axis of which corresponds to the temperature of a quantity of air (at any pressure and in a container of any volume) and the vertical axis of which is the absolute humidity of the air.

Step 2 Inscribe on such a grid a dot that shows the maximum value of absolute humidity of air that is at 70°F and standard atmospheric pressure.

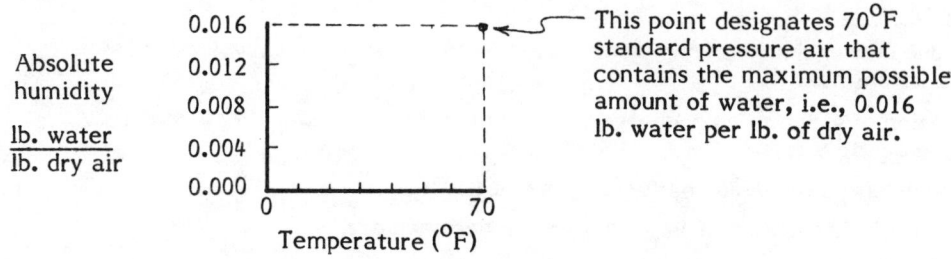

Assumption concerning pressure: In all subsequent graphs it is assumed that the air in question is at standard atmospheric pressure and remains so irrespective of any changes in temperature and humidity.

Step 3 Inscribe a curve showing, for every choice of temperature of the air in question, the maximum possible value of absolute humidity. The curve is called the saturation temperature curve or 100% RH curve. Higher absolute humidity values than indicated by this curve cannot be achieved; if someone tried to disperse more water into the air, so as to imply a point lying above this curve, he would fail.

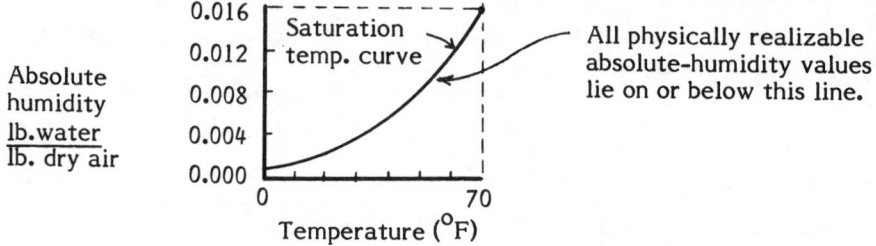

Step 4 Consider the following set of curves. The uppermost curve, as in the preceding graph, shows the maximum amount of water that can be included in standard-pressure air at any given temperature; it is called the saturation temperature curve, or 100% RH curve. The next curve shows, for every temperature in the indicated range, 90% of the maximum amount of water; it is called the 90% RH curve. Another curve shows 80% of the maximum amount. And so on. There are ten curves in all and they show 100%, 90%, 80%, 70%, 60%, 50%, 40%, 30%, 20%, and 10% of the maximum possible amount of water in standard-pressure air at any specified temperature in the indicated range.

57

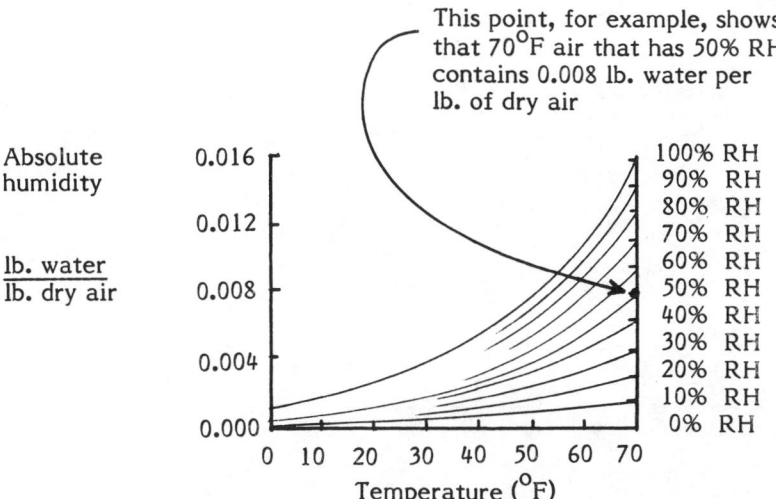

This point, for example, shows that 70°F air that has 50% RH contains 0.008 lb. water per lb. of dry air

Note that each curve is obtained merely by scaling down the 100% curve. That is, the curves are identical except for "vertical compression". Once someone has found, by experiment, exactly how to draw the 100% RH curve, the rest of the curves are found from it by simple arithmetic, i.e., by applying the multipliers 0.9, 0.8, 0.7, etc.

Step 5 Move the vertical scale to the right side of the graph. Why? Because the most interesting portions of the curves are the right portions; this is "where the action is". To save space, abbreviate the legends, units, etc., or inscribe them within the graph proper. (In most texts, the abbreviation process is carried too far! A reader lacking 20-20 vision and well-developed intuition may be at sea!)

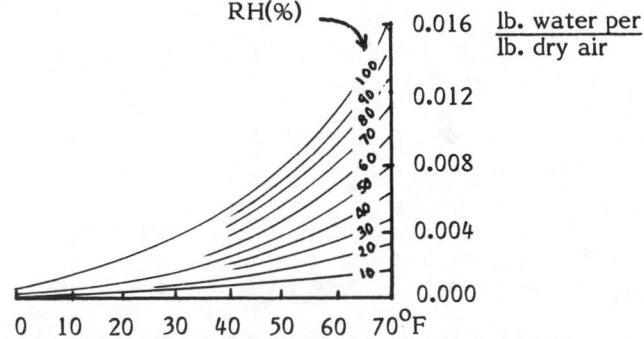

Step 6 Draw a line that designates all the temperature-and-absolute-humidity conditions that have a wet-bulb reading of 58°F. (Wet-bulb reading has been defined in Chapter 4. It is not to be confused with saturation temperature. Air that is at 70°F and includes 0.0078 lb. water per lb. of dry air has a wet-bulb reading of 58°F and a saturation temperature of 49°F.)

Some experimental facts: The wet-bulb reading will be 58°F if

the air in question is at 58°F and the RH is 100%,
the air in question is at 70°F and the RH is 50%,
the air in question is at 90°F and the RH is 10%.

Every point on the above-specified line represents air the properties of which are such that the wet-bulb reading is 58°F.

The added curve shown on the following graph designates all conditions for standard pressure air at 0 to 90°F that have a wet-bulb reading of 58°F. Notice that the curve is practically a straight line and has an upward-to-the-left slope. The curve is drawn on the basis of experimental data obtained many decades ago. Thus it represents, or summarizes, experimental facts.

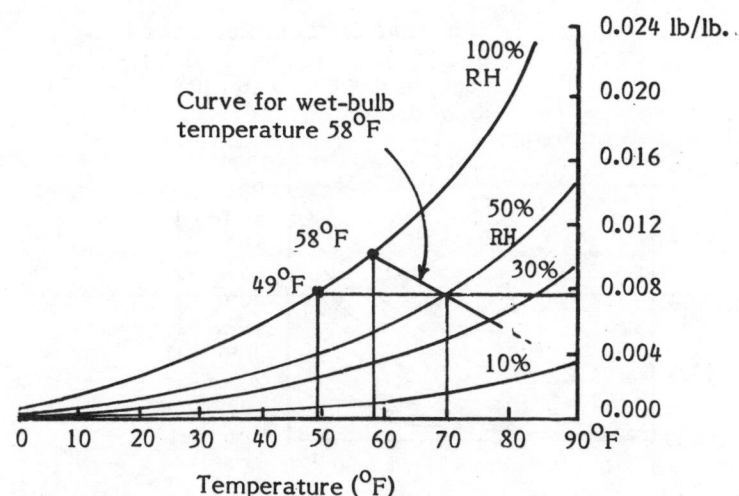

Temperature (°F)

Step 7 Add corresponding curves for every exact degree of wet-bulb reading, e.g., for wet-bulb readings of 58°F, 59°F, 60°F, etc. Using this set of curves (each of which is a nearly-straight line that slopes upward to the left), one can relate every combination of air temperature and absolute humidity to a wet-bulb reading, and vice versa. More exactly:

Given any two of these quantities:

      air temperature
      absolute humidity
      relative humidity
      wet-bulb reading,

-- an investigator can at once evaluate the other two, merely by inspecting the graph.
As before, standard pressure is assumed.

A formal, detailed graph is presented on a following page.

Step 8 Add, at any convenient location in the left portion of the graph, a straight line that has a certain slope and a certain scale, and call this the enthalpy scale. (As explained in Chapter 9, enthalpy is the name for the combination of two kinds of energy: heat energy (such as is possessed by any mass of gas irrespective of its pressure and volume) and "pressure-volume" energy (such as is possessed by any quantity of gas irrespective of its heat content).) The usual unit of enthalpy is: Btu per lb. of dry air.

Add, also, a set of straight parallel guide lines that are nearly perpendicular to the enthalpy scale, departing from perpendicularity by just the right amount.

If these additions are correctly made, the enthalpy of a pound of standard-pressure air can be ascertained almost instantly -- if any two of the above-listed quantities are known. One merely locates the pertinent point on the graph (say, the 70°F and 50% RH point), and, using the guide lines, finds the pertinent point on the enthalpy scale (the 26-Btu point, in the present example).

The enthalpy scale refers to an arbitrary zero: the enthalpy of 0°F totally dry air is arbitrarily called zero.

The procedures used in finding the correct slope and scale for the enthalpy line are outside the scope of this book. The same applies to the procedure for finding the correct slope of the guide lines. The procedures depend, of course, on the choices of scale for temperature and scale for absolute humidity. In the final version of the multi-purpose humidity graph shown on the following pages, these latter choices are such that the enthalpy scale is at about 46 deg. from the horizontal axis (temperature axis).

Annotated Multi-Purpose Humidity Graph
(Psychrometric Chart) for standard pressure air

Examples of use of enthalpy scale:

To convert 1 lb. of air from A to B (i.e., to heat
the air without changing its absolute humidity)
requires, as reference to the enthalpy scale
shows, the addition of about 17 Btu.

To convert from B to C (i.e., to increase the
absolute humidity while keeping the temperature
constant) requires about 17 Btu. (Same
amount -- by coincidence).

To convert from C to D requires no enthalpy!
The gain in sensible heat just equals
the loss in latent heat.

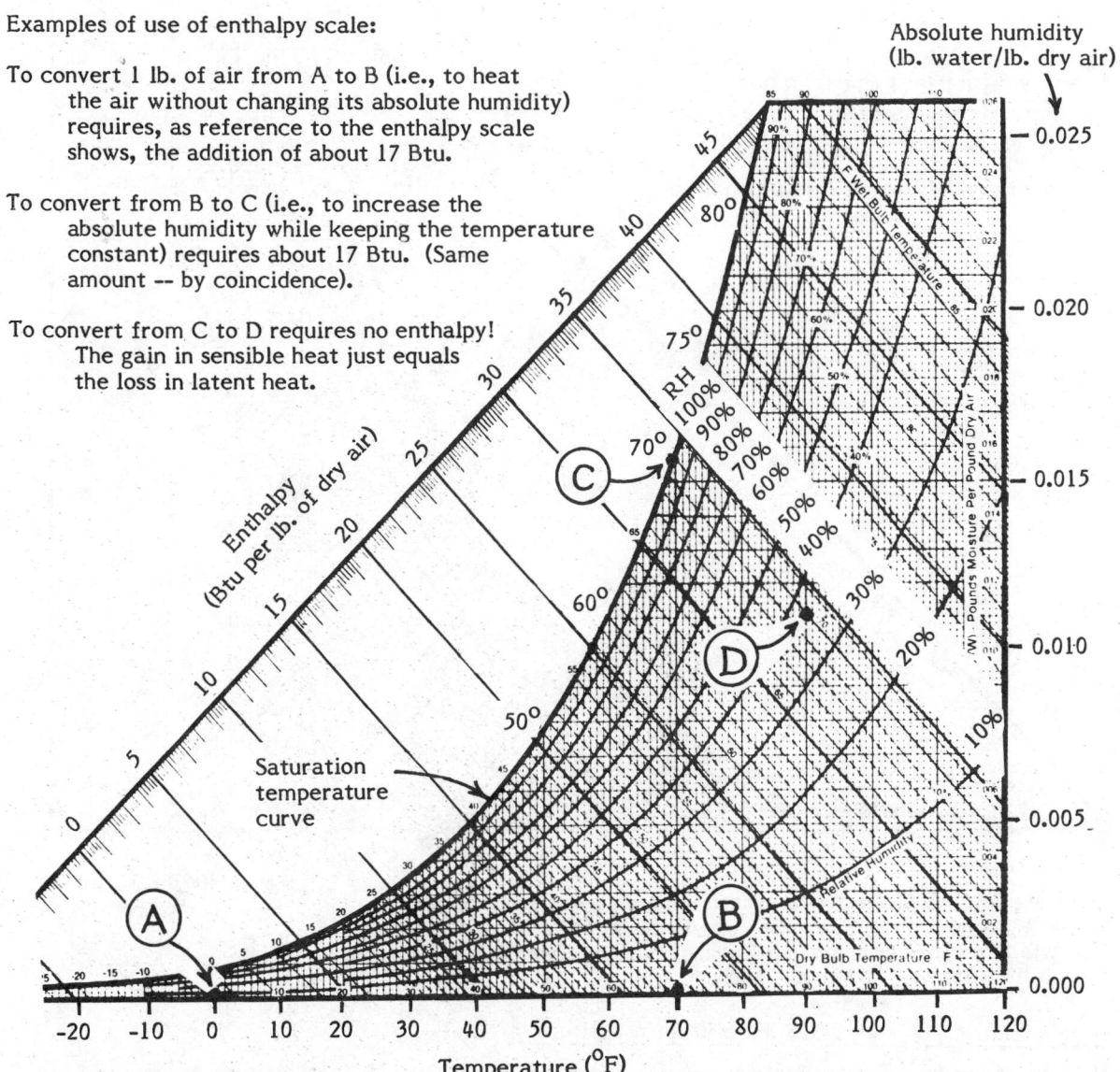

Saturation
temperature
curve

Temperature (°F)

Absolute humidity
(lb. water/lb. dry air)

Warning

Typical presentations of psychrometric charts have been slightly "fudged". If one wishes to be
strictly accurate in preparing such a chart, one can arrange for no more than two parameters to
conform to straight and uniformly spaced lines. Often, however, the draftsman shows additional
sets of straight and uniformly spaced lines: this makes his task easier and makes the graph easier
for persons to understand. The price paid is a very small sacrifice in accuracy -- all as explained
in Bibl. item S-162.

Note: The chart contains two set of lines that slope upward to the left: set of guide lines pertinent
to the enthalpy scale (lines of constant enthalpy), and set of constant wet-bulb-temperature lines.
Lines of the former set are straight and parallel. Lines of the latter set are not quite straight and
not quite parallel. All of which may produce some "visual confusion".
    Presented below are (1) an extra-clear, annotated, tutorial-version of the multi-purpose
humidity graph, or psychrometric chart, and (2) a pristine graph.

60

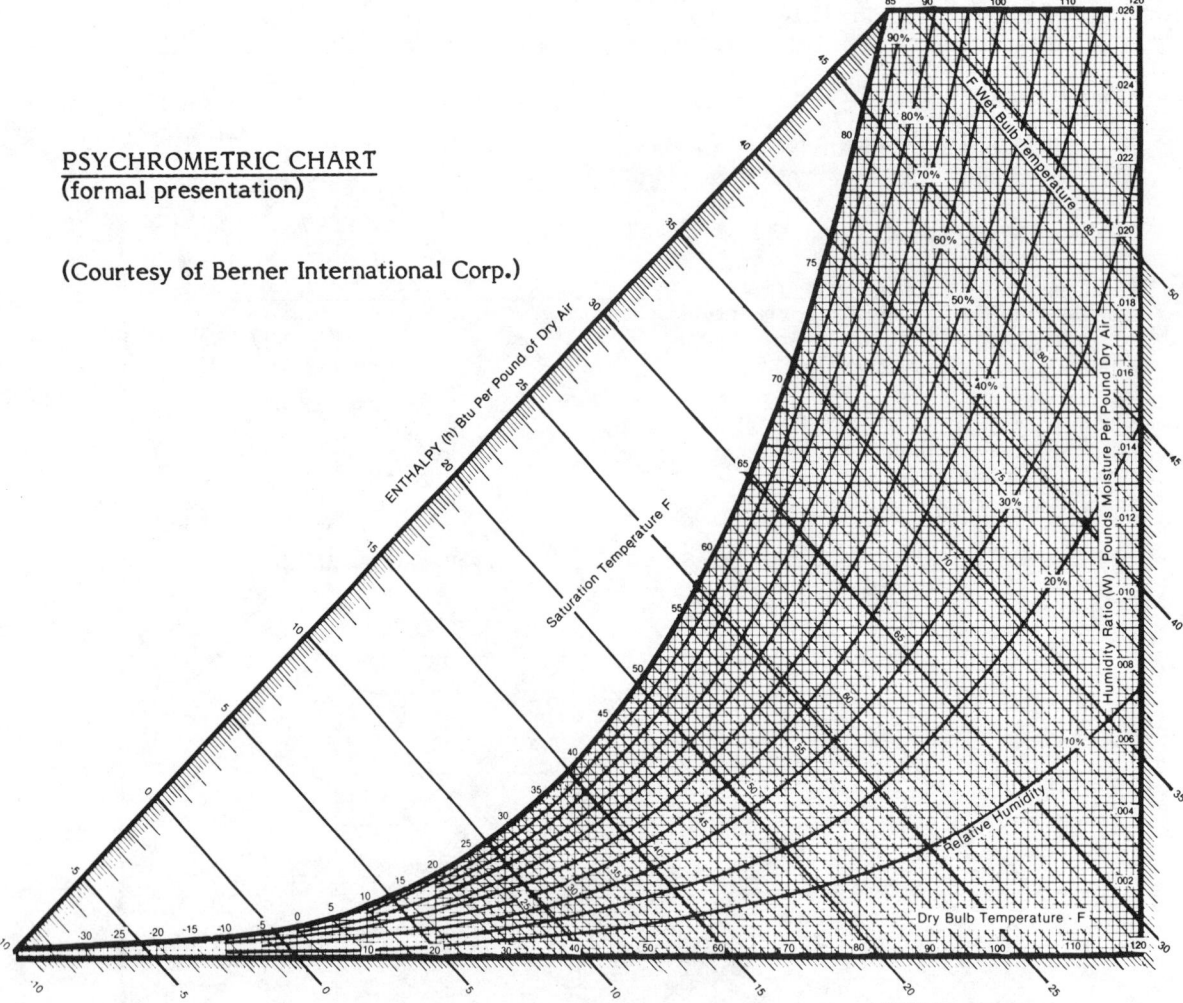

PSYCHROMETRIC CHART
(formal presentation)

(Courtesy of Berner International Corp.)

Knowing any two of these quantities: air temperature, absolute
humidity, relative humidity, wet-bulb temperature, enthalpy, anyone familiar with this chart can quickly find the other three quantities.
Hence the enormous power of the chart.

Psychrometric Properties Of Air At 5000 Ft. Altitude

Millions of persons in our western states live in towns and cities at 5000-ft. altitude or above. At
5000 ft. above sea level, atmospheric pressure is about 83% the pressure at sea level and the air
density is about 86% the density of sea-level air. A significant fact is that the higher the altitude
(and the lower the pressure), the greater the amount of water vapor relative to the amount of dry
air. For example, 70°F 50% RH air at sea level contains about 0.008 lb. of water per lb. of dry air
whereas 70°F 50% RH air at 5000 ft. altitude contains 0.0095 lb. of water per lb. of dry air. The
values of enthalpy per pound of dry air in these two cases are 25.5 Btu (at sea level) and 27 Btu (at
5000 ft.); the latter value is higher because of the greater amount of water vapor; more water vapor
means more latent heat, more enthalpy. For higher temperatures and higher relative humidities,
the disparity becomes even greater; values of absolute mass of water per pound of dry air and values
of enthalpy may be 10 to 20% greater than for sea-level air. Of course, the enthalpy values pertinent
to 0% RH air (bone dry air) at high altitude are the same as those for sea-level air; in both cases
there is no latent heat. Some handbooks and some manufacturers' brochures present a variety of
psychrometric charts, one for each choice of altitude, such as sea level, 2500 ft., 5000 ft., 7500 ft.

APPLICATION OF GRAPH TO AN EXCHANGER THAT EMPLOYS HEAT-TRANSFER SHEETS THAT ARE IMPERMEABLE TO WATER

Case 1: Heat Exchange Without Condensation

Consider the points A and B on the following graph. They represent quantities of dry (0% RH) air that are at 30°F and 65°F respectively. Projecting these points upward to the left onto the enthalpy scale, one obtains the values 7½ and 16 Btu per lb. of dry air. This means that to heat a pound of dry air from 30°F to 65°F takes 16 - 7½ = 8½ Btu.

Now consider the points C and D, representing quantities of air that have an absolute humidity of 0.003 lb. water per lb. of dry air and are at 70°F and 35°F respectively. How much enthalpy does a pound of such 70°F air give out when cooling to 35°F? Projecting points C and D onto the enthalpy scale, one arrives at the values 20½ and 12. The difference is 8½ Btu.

Comparing the results obtained in these two paragraphs, one sees that the enthalpy changes are equal and opposite. This means that a pound of dry 30°F air that is to be heated to 65°F requires exactly the amount of energy (enthalpy) that is given out by a pound of 70°F air (with absolute humidity 0.003) cooling to 35°F. That is, the 35-degree warm-up of the former air requires exactly the energy surrendered by the 35-degree cool-down of the latter air.

It is apparent from the graph that the same conclusion applies for a wide range of values of absolute humidity. Each airstream may have any absolute humidity (within the limit discussed below), and the same conclusion applies: there is complete energy match (enthalpy match) if the two temperature changes are equal and opposite.

This conclusion seems entirely obvious. But, in a sense, it is the graph that demonstrates the validity of the conclusion.

The conclusion is true as long as there is no condensation of water, and no evaporation of water, in either airstream. It is true, by the way, irrespective of whether the exchanger's efficiency of sensible-heat transfer is high or low.

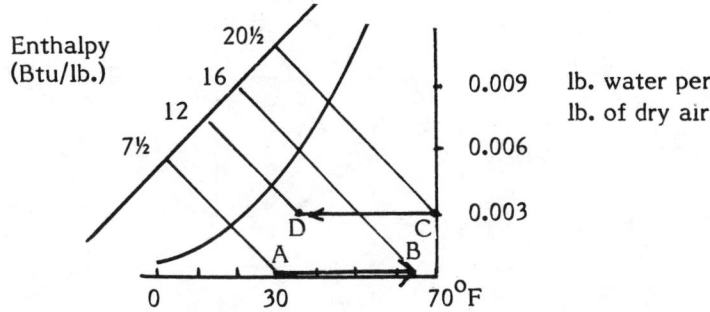

Case 2: Heat Exchange With Condensation

Suppose that the room air is very humid: suppose it is at 70°F and 70% RH (see point C on the following graph). How much must one pound of this air cool down in order to supply the heat needed to warm one pound of the dry outdoor air from 30°F to 65°F?

Note that the second half of the question (the outdoor-air story) is the same as in the case considered above, where it was shown that the warm-up process requires 8½ Btu. Accordingly, the outgoing air must supply 8½ Btu. Because the warm air has, at the start, 29 Btu, it must have, finally, 8½ Btu less, i.e., 20½ Btu (of enthalpy).

But, alas, there is no point on the horizontal line pertinent to 70°F and 70% RH that has an enthalpy as low as 20½! Such a low enthalpy cannot be achieved until (a) the RH of the warm air has reached 100% and (b) much water has condensed from this air. One must run his eye to the left along the horizontal line -- until the saturation temperature curve is reached; and one must then run his eye downward to the left along this curve until the point corresponding to enthalpy 20½ is reached. That point corresponds to 50°F, 100% RH. See point D.

Thus the answer is: To warm up the incoming dry air from 30°F to 65°F, the outgoing air must cool down from 70°F to 50°F, and it does this in two steps: (1) it cools with no change in absolute humidity -- cools until the RH becomes 100%, then (2) it cools further while water continually condenses from it and the RH remains at 100%. These two steps are represented by the two segments of the curve CD.

It may at first seem surprising that, to heat up the cold air 35 degrees, the warm air cools down only 20 degrees! The two temperature intervals do not match!

The reason they do not match is, of course, that the air that is cooling down is losing energy in two ways: by loss of sensible heat and by loss of pressure-volume energy. What is essential is that the two overall energy changes match. The two temperature changes do not have to match. There is a law of conservation of energy. There is no law of conservation of temperature-change.

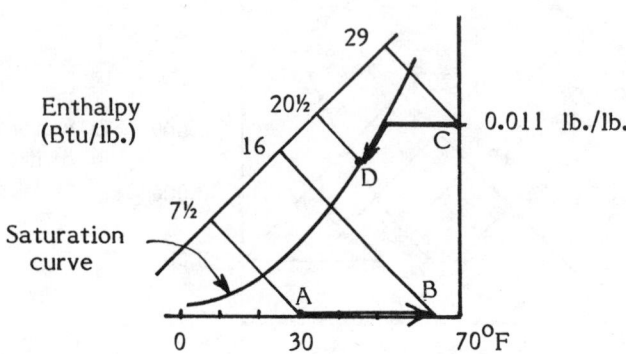

Paradox Concerning Latent Heat Of Water

In reading the preceding paragraphs, someone may feel that something has been omitted. He might say: "Wait! You have overlooked something! When the air has cooled so much that condensation occurs, something very special is happening: gaseous water is changing to liquid water. Accordingly the latent heat of water, 1054 Btu/lb., enters the picture. The graph refers to air -- not to liquid water or a combination of air and liquid water. Thus we have a new contribution of energy: the latent heat. Yet the previous discussion makes no mention of latent heat!"

Such objection is, in fact, not valid. The graph already takes fully into account the loss of enthalpy through loss of gaseous water. The saturation temperature curve falls very steeply largely because of the loss of gaseous water.

A key point is that, as far as the graph is concerned, it is irrelevant whether gaseous water is lost invisibly, as a gas, or is lost visibly as newly formed water droplets. Specifically: (1) If it lost invisibly, as a gas, it is gone and it carried away much enthalpy with it. (2) If it is lost visibly, as droplets, it is gone and has delivered much enthalpy (latent heat, to be more specific) to whatever other systems provided the cold surface on which the droplets formed. Either way, the water is gone and much enthalpy is gone. (The real difference is that in the former case the other system received much latent energy and little heat, whereas in the latter case it received less latent energy and more heat. It is a little like my handing you a check for $5 ("latent money") or a $5 bill ("sensible money"). Either way, I am $5 poorer and you are $5 richer.

If you are dying of thirst and want a glass of water, a heat-exchange process that yields a glass full of condensate is utterly different from one that delivers a lot of 99% RH air and no condensate. Yet the two processes may entail the identical decrease in enthalpy.

Warning: The graph takes no cognizance of what becomes of the water. If the water itself later cools off, giving its heat to another system, this is outside the scope of the graph. Usually the amount of heat that the water itself gives off, in cooling a few degrees, is negligible. If it is not, a separate calculation is required.

What Becomes Of The Latent Heat That Is Lost From The Outgoing Air When Condensation Occurs?

This heat is deposited on the surface that produced the condensation, i.e., on a cold surface, specifically a surface that is kept cold through its contact with incoming air. Clearly, then, the heat in question is received by the incoming airstream. This is, of course, very welcome (in winter).

What Becomes Of The Water Itself?

Inasmuch as we have been assuming that the heat-exchange surfaces are impermeable to water, it is clear that the water will not join the incoming air. Presumably it will drip onto the ground or flow into some drainpipe. (Most exchangers are designed in such a way that any such water will find its way to a drain; usually a 10-ft-long plastic drainpipe is provided.)

REMINDER THAT EXCHANGERS EMPLOYING IMPERMEABLE SHEETS COMMONLY DO RECOVER SOME LATENT HEAT

One might suppose that, inasmuch as latent heat is attributable to water vapor, an exchanger that recovers no water vapor can recover no latent heat.

This is not true. It is simplicity itself to recover latent heat without recovering water. It happens all the time -- in exchangers in which condensation of water is occurring. The condensation occurs on the surfaces that are made especially cold by the incoming cold air, and accordingly the heat liberated in the condensation process is imparted to the (impermeable) heat-transfer sheets and travels through them and warms the incoming air. In the extreme case where 70°F room air has an RH close to 100%, latent heat recovery may far exceed sensible heat recovery!

Obviously, if the condensation occurs -- not in the exchanger -- but in the outdoors, e.g., on the ground or on the outer surface of the foundation walls, the newly released sensible heat does not find its way into the stream of incoming air.

Warning That "Sensible-Heat Exchanger" Is A Deceptive Name

Condensation can occur in any heat-exchanger (if the outdoor temperature is very low) irrespective of whether the heat-transfer surfaces are permeable or impermeable. Thus any statement such as "This exchanger recovers sensible heat only", or "This exchanger does not recover latent heat" is wrong.

What is meant, typically, is: "As long as conditions are such that no condensation occurs, this exchanger recovers sensible heat only."

Important Bonus

The efficiency of heat-recovery by an exchanger that does not recover moisture is routinely evaluated under conditions that do not lead to condensation. Yet in especially cold weather much condensation does occur and thus much latent heat is recovered. This may be regarded as a bonus -- a very welcome bonus inasmuch as it is received just when most needed, i.e., when the outdoor temperature is low. It constitutes heat recovery over and above what the formal specifications imply.

I am indebted to Greg Allen for calling these matters to my attention.

APPLICATION OF GRAPH TO AN EXCHANGER THAT EMPLOYS HEAT-TRANSFER SHEETS THAT ARE PERMEABLE TO WATER

Case 1: Heat Exchange Without Condensation

Suppose the outdoor air is at 30°F and 0% RH and, on passing through an exchanger that has water-permeable sheets, gains heat and water and ends up at 65°F and 30% RH. Suppose the indoor air is at 70°F and 50% RH. What must the final condition of the outgoing air be, to supply the amount of heat and amount of water needed by the incoming air? The graph shows that the incoming air starts with an enthalpy of 7½ and ends with 20 (see points A and B on graph shown below). Thus it has gained 12½ Btu. The graph shows also that this air has gained 0.004 units of absolute humidity (0.004 lb. per lb. of dry air).

The outgoing air starts with an enthalpy of 25 (see point C). If it is to lose 12½ Btu (to provide match), it must end up with an enthalpy of 25 - 12½ = 12½ Btu. The outgoing air starts, also, with an absolute humidity of 0.0075: but to supply 0.004 to the other airstream it must end up with 0.0035.

What point on the graph corresponds to enthalpy 12½ and absolute humidity 0.0035? The point that has temperature 35°F and 80% RH (see point D). This is the answer.

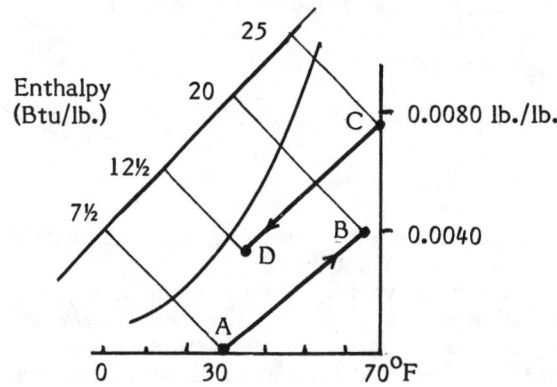

Formal statement of procedure for such case: In a case such as that discussed above, the steps that one takes in finding the properties of the outgoing air as it reaches the outdoors as follows:

1 Knowing the temperature and RH of the outdoor air, find (from the graph) the absolute humidity and enthalpy of this air.

2 Knowing the temperature and RH of the incoming air as it enters the room, find (from the graph) the absolute humidity and enthalpy of this air.

3 Compute, by simple arithmetic, the change in absolute humidity and change in enthalpy of the incoming air.

4 Compute the changes in absolute humidity and enthalpy in the outgoing air: the changes are necessarily the negatives of the above-mentioned changes.

5 Knowing the temperature and RH of room air, find (from the graph) the absolute humidity and enthalpy of this air.

6 Compute, by simple arithmetic, the absolute humidity and enthalpy of the outgoing air as it reaches the outdoors: simply apply (to the Step 5 data) the changes computed in Step 4.

7 Find, by referring to the graph, the temperature and RH of this air.

If the exchanger sheets are impermeable, i.e., if there is no transfer of water, the procedure is simpler. The temperature decrease of the outgoing air is equal to the temperature increase of the incoming air.

Converse process: Obviously the procedure discussed above can be reversed, so that a person who knows the properties of the outgoing air (before and after it passes through the exchanger) and of the outdoor air can compute the properties of the incoming air entering the room.

Case 2: Heat Exchange With Condensation

Suppose that the outdoor air condition is the same as in the previous case, but that the room air is much more humid: assume that it is at 70°F and 80% RH, which implies an enthalpy of 31 Btu and an absolute humidity of 0.0126.

If the room air, on passing through the exchanger, is to transfer the postulated 12½ Btu of enthalpy and the postulated 0.004 units of absolute humidity to the incoming air, the outgoing air must surrender 12½ Btu and surrender 0.004 units of absolute humidity. This implies that, on reaching the outdoors, this air must have an enthalpy of 31 - 12½ = 18½ Btu and must have an absolute humidity of 0.0126 - 0.004 = 0.0086.

There is no such point on the graph! There is no possible change in the room air that can deliver just the specified amount of enthalpy and specified amount of water -- unless some condensation occurs.

What must actually happen, to produce the postulated changes in the incoming air, is this: the outgoing air cools (and transfers gaseous water to the other airstream) until the RH of the outgoing air reaches 100%, and then two things happen: (1) water condenses from this air -- actually forms droplets -- and (2) this air cools further -- cools until it reaches the saturation-temperature-curve point (namely 47°F, 100% RH) that corresponds to the enthalpy value 18½ Btu.

66

The final state of the outgoing <u>air and water</u> is:

The air is at 47°F and 100% RH.

The water droplets are at 47°F or somewhat higher temperature. (Where the droplets go, and how much heat flows to or from them, are topics outside the scope of this discussion.)

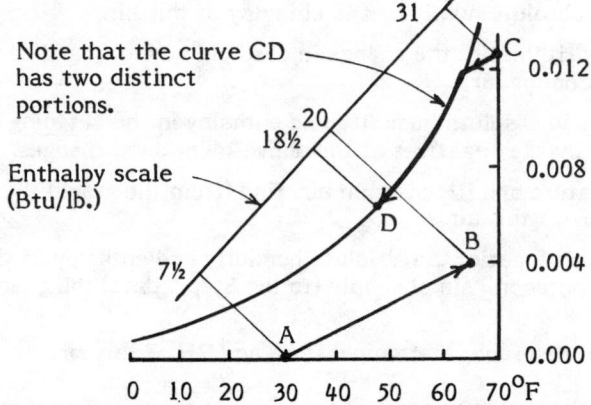

The curve CD has two segments: an upper segment that slopes slightly downward to the left and does not entail any condensation, and a lower segment that is part of the saturation temperature curve and slopes steeply and entails much condensation, i.e., formation of water droplets.

How much water is there in these droplets? This is easily computed. A pound of the postulated room air contains 0.0126 lb. of water. A pound of the air that has reached the outdoors (and is at 47°F and is saturated) contains 0.0069 lb. of water. And a pound of the postulated incoming air that is entering the room contains 0.004 lb. of water. Thus the amount of water in the droplets (per pound of room air involved) in this illustrative case is:

$$(0.0126) - (0.0069 + 0.0040) = (0.0126) - (0.0109) = 0.0017 \text{ lb.}$$

This is about 1/7 of the amount in one pound of room air.

As explained on a previous page, any changes that take place in the liquid water -- any changes in temperature and enthalpy -- are <u>not</u> represented in the graph. If the changes are of interest, they must be calculated separately. Usually the change in enthalpy of this quantity of liquid water is trivial; for example, in the present case the amount of liquid water is only 0.0017 lb., and accordingly if this amount were to change temperature by 10 F degrees the change in enthalpy would be only 0.017 Btu.

<u>What Becomes Of The Liquid Water?</u>

Inasmuch as we have assumed, in this case, that the heat exchanger surfaces are permeable to water, it is possible that a fraction of the water will find its way to the stream of incoming air and will evaporate into that air.

Such evaporation will have a cooling effect. In winter, cooling effect is undesirable. Accordingly the designer may design the exchanger so that most of the condensate will drip onto the ground or into a drainpipe.

In some circumstances (e.g., very cold day in winter), much of the liquid water will freeze. Or perhaps, initially, it will condense into frost rather than liquid water. Frost can, of course, cause trouble, as indicated in other sections.

Some Assorted Reminders, Warnings, And Confessions

Usually I am dealing with one pound of air. If a different quantity is involved, the results must, obviously, be scaled up or down proportionally.

Sometimes I deal with one pound of <u>dry</u> air and sometimes one pound of ordinary air. I am inconsistent. However, the discrepancies are small, of the order of 1%, inasmuch as the water content of air is, typically, only about 1% (by weight) of the air as a whole.

Nearly always I assume that the airstreams have the identical flowrate. In some exchangers the two flowrates may be adjusted so as to be very different, for example, they may be in the ratio of 3 to 2. In such cases the necessary calculations are, of course, different.

"Identical flowrates" can have two different meanings: same <u>volume</u> flow in each stream or same <u>mass</u> flow. That is, equality can be specified in terms of cfm or lb/min. In those calculations in which I am inconsistent in this respect, errors of the order of, say, 2 to 8% may be incurred. However, such errors may be acceptable in view of the many uncertainties elsewhere in the system -- uncertainties as to the line voltage and power supplied to the blowers, pneumatic resistance of the exchanger proper and its manifolds and ducts, and effects of dust and water droplets or frost that may have accumulated within the exchanger proper.

Usually I ignore the enthalpy changes in any already-condensed water.

Usually I assume that the exchanger includes no frost or ice.

Some of the sets of outcomes that one may derive from the graph -- for various assumptions as to initial conditions of the airstreams -- are outcomes "in principle". That is, they are outcomes that can be achieved if a sufficiently well-made exchanger is available. In other words, they are <u>possible</u> outcomes that <u>can be achieved</u> if one is willing to go to unlimited trouble and expense.

The discussions pay little or no heed to the second law of thermodynamics. Yet that law is very important. It rules out a great variety of seemingly acceptable performances. For example: Suppose Smith says: "I intend to make an exchanger (with water-impermeable sheets) that will cool the outgoing air from 70°F to 0°F and heat the incoming air from 30°F to 100°F." This proposal may seem reasonable to anyone familiar only with the first law of thermodynamics: law of conservation of energy (enthalpy). The proposed performance conforms to this law: the amount of energy given up by the outgoing air exactly equals the amount received by the incoming air. The second law, however, knocks the scheme on the head: it declares it to be impossible, within a closed system, to decrease the entropy, i.e., to make a net increase in the usefulness of the energy, such as would be the case if, starting with equal quantities of 30°F and 70°F air, one were to end up with equal quantities of 0°F and 100°F air. If one could do that, one could soon devise a perpetual motion machine.

CORRESPONDENCES BETWEEN POINTS ALONG THE TWO CURVES

It is moderately interesting to see the relationship between points along the two curves on the multipurpose humidity graph -- points that correspond to positions along the two airstreams in a counterflow heat-exchanger.

Suppose that the exchanger proper is 16 in. long. Consider points that are 0, 4, 8, 12, and 16 inches from the left (cold) end. What are the properties of the air in each stream at these locations along the exchanger? At each location, how do the properties of the two streams differ?

The following graph pertains to a moderately efficient exchanger that employs water-impermeable heat-transfer sheets.

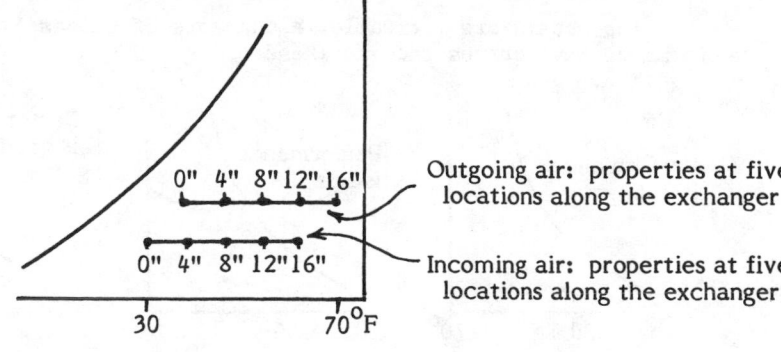

(All locations are in inches from the cold end.)

68

Another example: Suppose the exchanger includes water-permeable sheets and is designed for very high all-around efficiency. Then the two curves and sets of illustrative points might have the form indicated below.

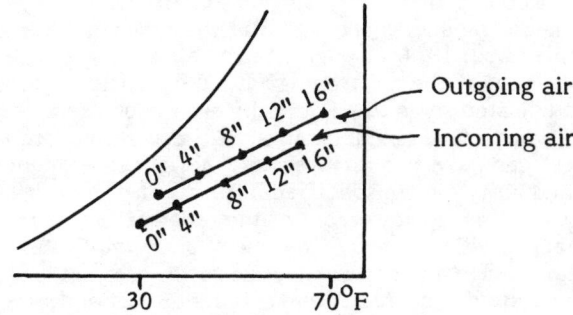

In Situations In Which No Condensation Occurs, Are The Two Curves Straight Lines?

If the transfer sheets are impermeable to water, yes: each curve is a straight horizontal line.

If the sheets are permeable, the answer is not so obvious. My guess is that, again, the two curves are straight (or near-straight) lines. My guess is based on the fact that the properties of the exchanger itself are approximately uniform (linear), and corresponding segments of the two curves must match.

What Can Be Said About The Sum Of The Two Enthalpy Values Pertinent To A Pair Of Points (For Example, The Two Points Pertinent To The Location That Is 4 in. From Cold End Of Exchanger)?

For each pair, the sum must have the same value. The law of conservation of energy requires this.

GRAPHICAL REPRESENTATION OF IMPROVEMENTS IN PERFORMANCE

How does the pair of curves change when one progressively improves the exchanger's performance?
If the sheets are impermeable, exchangers of successively improved performance have curves such as are sketched below.

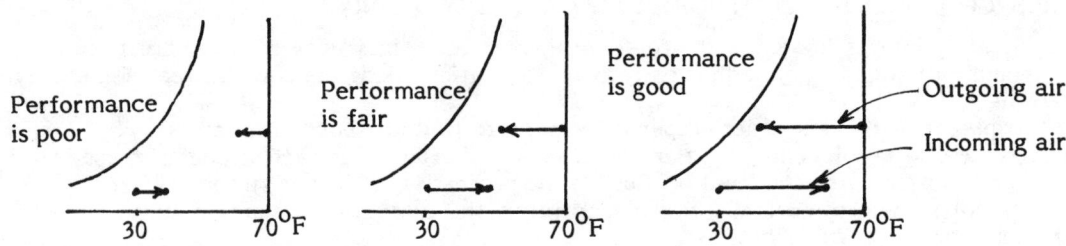

If the sheets are permeable, exchangers of successively improved performance have curves such as these:

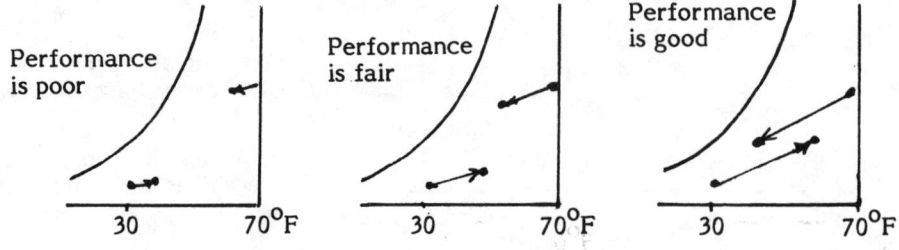

How do the two curves change if, in an exchanger that has mediocre performance in every respect, you improve <u>just the thermal conductivity</u>?

Answer: Each curve becomes more stretched-out horizontally. Each becomes longer and has a less steep slope, as suggested by the following graphs.

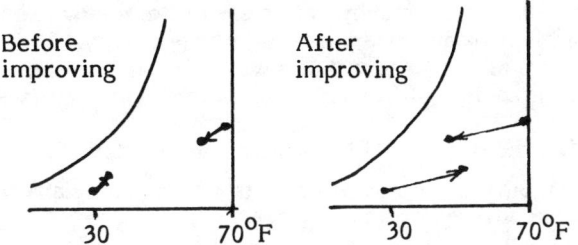

How do the two curves change if (without changing the thermal conductivity) you <u>improve the permeability to water</u>?

Answer: Each curve becomes steeper.

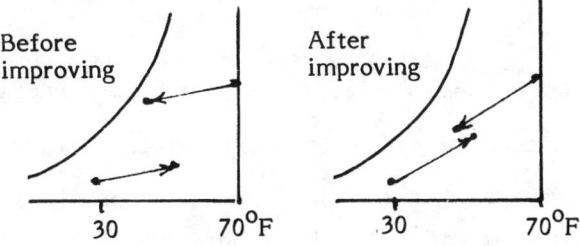

GRAPHICAL REPRESENTATION OF COLD-DAY VS. WARM-DAY PERFORMANCE

If the outdoor air is very cold and has very low absolute humidity, the two curves that describe the temperature changes in the two airstreams are longer and farther apart than is the case when the outdoor temperature is mild and has high absolute humidity. See following graphs, pertinent to an exchanger that transfers sensible heat only.

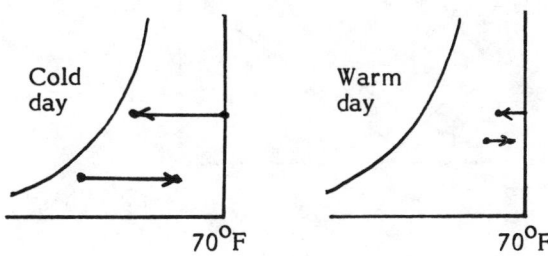

EFFICIENCY OF ENTHALPY EXCHANGE OF AN EXCHANGER THAT HAS WATER-PERMEABLE SHEETS

The simplest kind of efficiency -- efficiency of sensible-heat exchange--has been defined in Chapter 7.

Now I discuss a broader kind of efficiency: efficiency of enthalpy exchange, i.e., efficiency of saving, or recovering, sensible heat and also pressure-volume energy. This is the kind of efficiency of interest to designers and users of exchangers that have exchange-surfaces that are permeable to water. I call this kind of efficiency E_{enth}. Its definition is somewhat similar to that of efficiency of sensible heat exchange.

E_{enth} may be defined as the ratio of (1) the amount of enthalpy H that is recovered, thanks to the exchanger, from the outgoing air and transferred to the incoming air, to (2) the greatest conceivable amount that could be recovered by an exchanger.

Let H_1 be the enthalpy of a pound of room air (which is usually warm and has moderately high absolute humidity),

H_2 be the enthalpy of a pound of outdoor air (which, in winter, is usually cold and has low absolute humidity),

H_3 be the enthalpy of a pound of outgoing air as it reaches outdoors,

H_4 be the enthalpy of a pound of incoming air as it reaches the room.

Then $E_{enth} = \dfrac{H_4 - H_2}{H_1 - H_2}$.

(A practically equivalent definition is: $\dfrac{H_1 - H_3}{H_1 - H_2}$. If some condensation of water occurs, this definition may lead to questions concerning the temperature and disposal of the condensate.)

The following graph helps make the definition clear. Heavy bars at left represent $(H_4 - H_2)$ and $(H_1 - H_2)$, and E_{enth} is simply the ratio of the lengths of these lines; it is the ratio of the shorter to the longer.

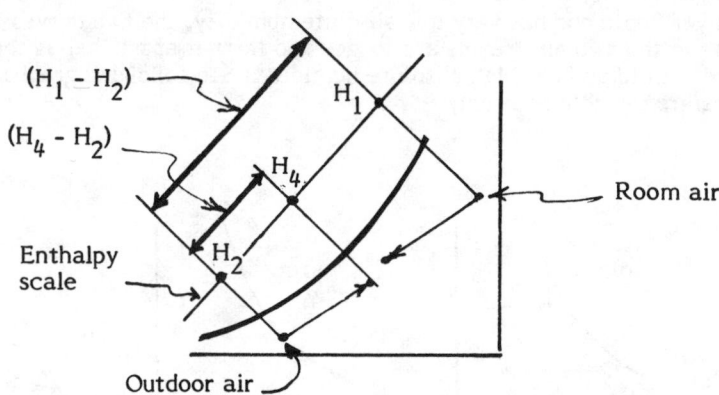

Example Suppose the enthalpy of the room air is 30 Btu/lb. and the enthalpy of the outgoing air as it reaches the outdoors is 20 Btu/lb. Suppose also that the enthalpy of outdoor air is 8 Btu/lb. and the enthalpy of the incoming air as it reaches the room is 18 Btu/lb. What is the efficiency of enthalpy exchange?

Here H_1, H_2, H_3, and H_4 are 30, 8, 20, and 18 Btu/lb respectively, and accordingly:

$$E_{enth} = \frac{H_4 - H_2}{H_1 - H_2} = \frac{18 - 8}{30 - 8} = \frac{10}{22} = 45\%.$$

Component Efficiencies

One can define a component efficiency of sensible heat exchange as the ratio of (a) actual amount of sensible heat transferred to (b) the maximum possible amount. To find the ratio in an actual case, one merely takes the ratio of the lengths of the lines L_1 and L_2 defined in the following sketch -- or one may use simple arithmetic.

Example If the incoming air rises 30 F degrees, but would have had to rise 40 F degrees to equal the room temperature, the component efficiency of sensible heat exchange is $30/40 = 75\%$.

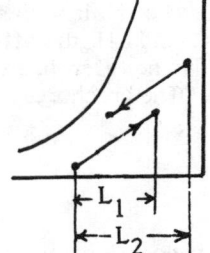

Likewise one can define a component efficiency of latent heat exchange as the ratio of the actual amount of latent heat transferred to the maximum possible amount -- i.e., as the ratio of lengths of the lines L_3 and L_4 defined in the following sketch.

Note This quantity might equally well be called efficiency of water recovery, or efficiency of moisture-in-air recovery.

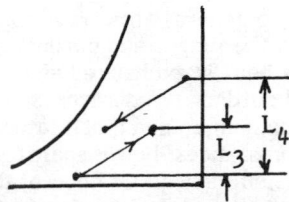

Obviously, the sum of the two component efficiencies may exceed 100%. For an ideal exchanger the sum would be 200%.

Note also that the average of the two component efficiencies may have little meaning. If, in a particular case, a large amount of sensible heat is available but only 10% of it is transferred, and a negligible amount of latent heat is available and 100% of it is transferred, the average of these efficiencies -- 55% -- has no meaning, inasmuch as the efficiency that counts here (efficiency of enthalpy exchange) is only about 10%.

Fractional Efficiencies

It is possible to define two fractional efficiencies, one pertaining to sensible heat and the other pertaining to latent heat, that add up to the overall (enthalpy exchange) efficiency. The sensible-heat fraction of efficiency of enthalpy exchange may be defined as (1) enthalpy recovery associated with the increase in sensible heat of the incoming air divided by (2) total enthalpy recovery. The latent-heat fraction of efficiency of enthalpy exchange may be defined as (1) enthalpy recovery associated with the increase in latent heat of the incoming air divided by (2) total enthalpy recovery.

I expect that these fractional efficiency concepts are of little use.

Warning: Enthalpy Exchange Efficiency Is Not A Constant Of An Exchanger

There is an important difference between (1) efficiency of sensible-heat exchange of an exchanger that has impermeable heat-transfer surfaces, and (2) efficiency of enthalpy exchange that has permeable heat transfer surfaces:

> The former efficiency is a constant of the exchanger (operated with the given rates of airflow). It does not change when the indoor temperature changes or the outdoor temperature changes. Even if these temperatures change considerably, the efficiency remains practically unchanged, provided no condensation occurs.

> The latter efficiency is not a constant. It depends on two different processes (transfer of sensible heat, transfer of latent heat), and the two pertinent component efficiencies are not equal, ordinarily. Thus if some change occurs in room air temperature but not humidity, or vice versa, a change will occur in the efficiency of enthalpy exchange. The same is true with respect to change in outdoor air temperature or humidity.

Accordingly it makes good sense to speak of "the efficiency" of an exchanger that exchanges sensible heat only, but it does not make good sense to speak of "the efficiency" of an exchanger that exchanges sensible heat and latent heat. For such an exchanger, one should say, rather: "When operating between room air with such-and-such temperature and RH and outdoor air with such-and-such temperature and RH, the efficiency of enthalpy exchange is such-and-such."

I have found no reference to this topic in the literature or in the manufacturers' brochures that I have seen. The brochures refer (impermissibly) to "the efficiency" of enthalpy exchange.

IN ENTHALPY EXCHANGE, WHICH IS LARGER: SENSIBLE HEAT EXCHANGE OR LATENT HEAT EXCHANGE?

There is no general answer. Depending on circumstances, either can be larger. For example, on a cold day when the indoor and outdoor absolute humidities happen to be the same (a rare situation, obviously), no latent heat is exchanged and much sensible heat is exchanged. Conversely, on a day when the indoor and outdoor temperatures happen to be the same but the absolute humidities differ, no sensible heat is exchanged and much latent heat is exchanged. In summary, the relative magnitudes depend on the circumstances: how disparate are the indoor and outdoor temperatures and the indoor and outdoor absolute humidities. Of course, the characteristics of the exchanger itself are pertinent also.

"Which is larger typically, assuming that the exchanger is equally effective with respect to sensible heat and latent heat?", one may ask. Some illustrative answers are as follows -- for situations where the indoor and outdoor humidities are both 50%:

> Outdoor and indoor temperatures $0^\circ F$ and $70^\circ F$:

>> Latent heat exchange is 56% as large as sensible heat exchange. That is, latent heat recovery accounts for about 1/3 of the enthalpy recovery.

> Outdoor and indoor temperatures $30^\circ F$ and $70^\circ F$:

>> Latent heat exchange is 76% as large as sensible heat exchange.

> Outdoor and indoor temperatures $50^\circ F$ and $70^\circ F$:

>> Latent heat exchange is 100% as large as sensible heat exchange. Each accounts for half of the enthalpy recovery.

If the outdoor relative humidity were smaller (smaller than the above-assumed value of 50%), but the indoor humidity remains 50%, then, obviously, the relative magnitude of latent heat exchange will be greater than indicated above. And vice versa.

Additional generalization: In very cold and dry regions, the pertinent part of the psychrometric chart is the lower left part, and, here, the absolute amounts of water involved are small. Accordingly the potentiality for recovering latent heat is small. During a hot and humid spell in summer, the upper part of the chart applies, and, here, the amount of latent heat involved may be large.

UPPER LIMIT ON EFFICIENCY OF GASEOUS WATER TRANSFER

What is the greatest amount of gaseous water that can be transferred (through the water-permeable sheets of an exchanger) from the outgoing airstream to the incoming airstream when the exchanger is of ideal-maximum-efficiency type and the mass-airflows in the two airstreams are identical? Let us consider two cases: one that is simple but extreme, and one that is general.

Case 1

Countercurrent exchanger that has water-permeable heat-transfer sheets and has near-100% efficiency of enthalpy exchange. Room air RH is 100%, outdoor RH is 0%. The respective temperatures are 70°F and 0°F.

My guess is that it is possible in principle that the exchanger is so well designed and built that the fresh air, on entering the room, will be at almost 70°F and 100% RH, and the outgoing air, on reaching the outdoors, will be at almost 0°F and 0% RH. In other words there can be a "complete swapping" of conditions. I believe that application of the above-presented seven-step procedures bears this out, and neither the first nor second law of thermodynamics precludes it.

Case 2

As above, except that the room air specifications and outdoor air specifications can be any specifications.

My guess is that here, too, there can be complete swapping of the specifications.

Note: This latter case is very general indeed. It includes not only typical winter situations but also summer situations; that is, it includes cases where the warmest and highest-absolute-humidity air is the outdoor air.

A prohibition: My guess is that, if the outdoor air has lower temperature and absolute humidity than room air has, the incoming air that enters the room cannot have higher temperature or higher absolute humidity than the room air has. In other words, it is not possible for the two airstreams to "more than swap" specifications; I assume that the second law of thermodynamics prohibits this.

TWO METHODS OF TRANSFERRING GASEOUS WATER

The two well-known methods are:

Using a fixed-type exchanger, arrange for the water to pass from one airstream to the other: make the heat-transfer sheets of material that is permeable to water but not to air. In the Mitsubishi Lossnay exchanger the heat-transfer sheets are of specially treated paper that is permeable to water. Presumably various other porous or hydrophilic materials could be used.

Using a rotary-type exchanger, in which the two airstreams follow one another sequentially through the identical passages, impregnate the passage walls with a hygroscopic material, i.e., a desiccant. Lithium chloride has been used for this purpose. So has silica gel, which is non-deliquescent -- stays put despite long exposure to much water.

TWO KINDS OF CONVERSION OF RECOVERED LATENT HEAT

When latent heat is recovered from the outgoing air by virtue of condensation of water on water-impermeable surfaces cooled by the incoming air, the latent heat is converted to sensible heat. Thus it contributes to keeping room temperature high -- a desirable service in winter irrespective of room air RH.

When latent heat is recovered from the outgoing air by virtue of recovery of water (steady recovery through a water-permeable surface or intermittent recovery in a rotor that includes desiccant), the latent heat remains as latent heat. None of it is converted to sensible heat. None of it contributes to keeping room temperature high. It contributes only to room air humidity -- a

74

service that may be desirable or undesirable depending on whether the humidity tends to be too low or too high.

If a prospective purchaser is comparing exchangers some of which recover latent heat in addition to sensible heat, he should ask himself whether most of the latent heat recovered will be recovered merely at latent heat (increased humidity) and whether the increased humidity will be welcome.

NEAR-IMMUNITY-TO-CONDENSATION OF EXCHANGER THAT TRANSFERS WATER

From the various preceding graphs one sees a striking difference in the immunity-to-condensation of exchangers that do and do not transfer water. The difference may be summarized thus:

In an exchanger that has water-impermeable transfer surfaces, as the humid indoor air cools (in passing along with the exchanger), the characteristic point on the psychrometric chart moves straight to the left and hits the saturation curve -- indicating that condensation starts.

In an exchanger that has permeable transfer surfaces, the humid indoor air that travels along within the exchangers does two things at once: cools and loses water. The characteristic point moves downward and to the left and thus may miss the saturation curve entirely, or may strike it much farther along to the left. Accordingly condensation may not occur at all, or may occur only if the outdoor temperature is extremely low.

Example Suppose that room air is at 70°F and 50% RH and that the outdoor air is at 0°F and 0% RH. The path of the point describing the performance of a 100% efficient exchanger with water-impermeable transfer surfaces is shown by the line ABC in the following graph. Note that saturation occurs at point B and continues along segment BC. The path pertinent to a 100% efficient exchanger with water-permeable surfaces is shown by the line AD. Note that the line keeps clear of the saturation curve; no condensation occurs.

Of course, if the exchanger has a low efficiency of water transfer, the immunity under discussion is much reduced.

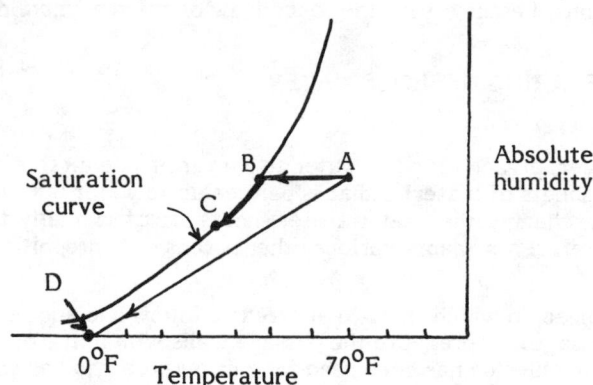

AT WHAT OUTDOOR TEMPERATURE DOES FROSTING OCCUR IN THE OUTGOING AIR OF AN EXCHANGER THAT HAS WATER IMPERMEABLE SURFACES?

In an exchanger that has water-impermeable heat-transfer surfaces (and has equal airflow rates in the two airstreams), frosting in the outgoing-air passages will not occur, ordinarily, if the outdoor temperature is only a little below 32°F. What usually occurs, when the outdoor temperature is in the range from 20°F to 32°F, is either (a) the outgoing air does not cool down to the pertinent saturation temperature and accordingly no condensation or frosting can occur, or (b) the cooling is great enough so that some condensation does occur, but the condensation process supplies enough sensible heat so that the outgoing air does not cool down as far as 32°F.

Example If the outdoor air is at 0°F and 0% RH and warms up to 50°F, it has gained about 12 Btu of enthalpy per pound. To supply this much enthalpy, when the indoor air is at 70°F and 50% RH, the indoor air cools (with some condensation) only to about 35°F; the cooling is modest because so much of the heat contributed is the contribution of latent heat (latent heat converted to sensible heat in the condensation process).

From examining the psychrometric chart, one finds, assuming an exchanger with water-impermeable surfaces and the highest possible efficiency of heat transfer, and assuming the room air to be at 70°F and a relative humidity of 30, 50, or 70%, and assuming any outdoor relative humidity whatsoever, that frosting in the outgoing-air passages can occur only if the outdoor temperature is at least as low as about 29°F, 16°F, or 3°F respectively.

If the room air is somewhat colder -- only 60°F, say -- the temperatures of frosting are considerably higher: for room air at 50% and 70% RH they are about 23°F and 14°F. If the room air is at only 30% RH, frosting will begin as soon as condensation begins, which will occur whenever the outdoor air is colder than 28°F; no liquid will be produced -- only frost.

If the room air is at 90°F and 50% RH the frosting temperature is much lower. Such air contains an especially large absolute content of water, the pertinent portion of the saturation curve is especially steep, the amount of condensate is large, the amount of latent heat converted to sensible heat is large, and accordingly the outgoing air will remain above 32°F even if the outdoor temperature is as low as -15°F.

Values Pertinent To A Less Efficient Exchanger

Let us assume that the exchanger efficiency is well below 100%: assume, specifically, that the rise in temperature of the incoming air is only 80% of the maximum possible rise (rise to equal the temperature of indoor air). Then the frosting temperatures are lower. The frosting temperatures applicable when 70°F room air is at 30%, 50%, or 70% RH are about 17°F, 3°F, and -15°F respectively.

Warning: The values presented above have been estimated by me from the psychrometric chart, and they differ moderately from the values presented on page 7.4 of the SMACNA book (Bibl. item S-162).

Note that the temperature of frosting decreases sharply when air that is hotter than 60°F and more humid than 50% is made even hotter or even more humid. That added warmth tends to suppress frosting is not surprising; but that added moisture also has this effect may at first sight seem surprising.

MEANS OF CONTROLLING FROSTING

If there is danger that frost will form in the outgoing-air passages of a fixed-type heat exchanger, these steps may be taken:

Preheat incoming air. Use small electrical pre-heater. Or (proposed by Advanced Idea Mechanics Ltd.) use indoor uninsulated fresh-air-intake duct from the outdoors to the exchanger.

Reduce flowrate of incoming air or increase flowrate of outgoing air.

Keep the room air at higher temperature and/or at higher relative humidity.

If the exchanger is of a type that does not transfer moisture, switch to a type that does.

Reducing a flowrate can be accomplished either by reducing the electric power supplied to the blower or by installing a damper that partly closes off the airstream.

Most of the above-listed operations can be performed manually or automatically with the aid of temperature of pressure sensors.

(Would it be possible to design an exchanger in which the added weight of a frost accumulation would deform the plate system in such a way that the passages for incoming air would be made smaller? If so, the desired reduction in rate of airflow of incoming air would be accomplished automatically, without the need for a special sensor and special actuator.)

Getting Rid Of Already Formed Frost

If frost has already accumulated in the outgoing-air passages of the exchanger, one may get rid of it by (1) turning off the incoming-air blower for a while, allowing the frost to melt and go away, or (2) gaining access to the cold end of the exchanger and delivering heat to it -- by means of hot or luke-warm water from a hose, or by means of an electrically powered hair-dryer, or by means of a special heating coil of some sort.

USE OF A DEHUMIDIFIER TO REMOVE AND/OR RECOVER WATER FROM ROOM AIR

If one wishes to reduce the amount of water in room air, he may use a commercially available device -- a dehumidifier -- that performs this task excellently. In such a device, an electrically powered cooling system keeps a large surface very cold, and water from the room air condenses on this surface.

One benefit of this procedure, in winter, is that each pound of water that condenses delivers to the dehumidifier (and thus to the room) 1054 Btu. Room temperature is increased not only by such 1054 Btu contribution but also by a portion of the electrical power expended in operating the dehumidifier. If electrical power is expensive, as it usually is, this may not be a cost-effective way of helping heat the house. However, if the house in question is already being heated by electrical power, bringing a dehumidifier into play in this manner may be highly cost-effective.

Another benefit is that the house occupant may put the recovered liquid water to use. If, at some time, room air is too dry, this water may be reintroduced into the room air, by natural evaporation or other means.

Solar Assisted Dehumidifiers

A recent report by SERI (U-457-HE-40) reviews various methods of dehumidifying air by employing a desiccant (to absorb the gaseous water from room air) and employing solar energy (to heat the desiccant and drive off the water -- so that the desiccant can be readied for re-use).

Note: Humidistats and dehumidistats are discussed in Chapter 16.

NEED FOR AVOIDING HIGH RELATIVE HUMIDITY IN HOUSES IN COLD CLIMATES

In midwinter, in a house that is in a very cold climate, allowing the relative humidity to become high (for example, exceed 50 or 60%) can be unfortunate. Much moisture may condense on windows, even if they are double-glazed, and may condense also on any poorly insulated walls, floors, etc. Under some circumstances even 35% RH may be too high: condensation may occur. In some of the newer, very-tightly-built houses, high humidity sometimes does occur.

In warm climates the problem is small. The need to avoid high RH is much less. (I am indebted to Greg Allen for telling me about this.)

SUMMER VS. WINTER USE OF EXCHANGERS

Obviously, summer use of exchangers may be of little importance in Canada and great importance in Florida. For most houses in USA the wintertime use is far more important than the summertime use.

In winter, when indoor absolute humidity is moderate and outdoor absolute humidity is very low, recovery of sensible heat is the main goal. The amount of latent heat that may be recovered is smaller -- slightly smaller if the outdoor temperatures are 30°F or 40°F, and smaller by a factor of 2 to 4 if the outdoor temperatures are about 0°F or -20°F.

In summer, when the outdoor air may be hot (90°F) and very humid, latent heat looms large; it may exceed the sensible heat. That is, excluding the outdoor moisture can pay big dividends.

The efficiency of sensible heat recovery is essentially the same, irrespective of whether the outdoor temperature is lower or higher than indoor temperature. As regards efficiency of recovery of latent heat, the situation is complicated by any condensation of moisture; I know of no interesting comparison of the efficiencies in winter and summer.

Chapter 12

ROTARY TYPE EXCHANGERS: DESIGN PRINCIPLES

INTRODUCTION

Rotary type exchangers have played a major role in industrial plants, commercial buildings, etc. They can handle very large airflows, yet occupy relatively little space. They can recover sensible heat and latent heat. Recovery efficiency is high.

But to designers and builders of houses, they are little-known. I myself was unaware -- until August of 1981 -- that rotary exchangers specially designed for household use existed.

Only one kind is available in the U.S.A., but my guess is that other exchanger manufacturers will soon develop competing models. The one kind now available, distributed by Berner International Corp., is made in Japan, where it is known as the Sharp Corp. Econofresher GV-120.

The important advantages of the rotary exchanger are discussed in this chapter, and comparisons with fixed-type exchangers are presented in Chapter 17. Among the advantages are: high recovery of latent heat (as well as sensible heat), relative freedom from condensation and frosting problems, and (especially in applications requiring large rates of airflow) compactness.

Excellent discussions of large-size, industrial-use, rotary exchangers are included in the book "Energy Recovery Equipment and Systems: Air-to-Air" published by the Sheet Metal and Air Conditioning Contractors National Assn.; see Bibliography item S-162.

BASIC PRINCIPLE

A rotary exchanger, which is one class of reversing flow exchanger, employs a slowly rotating wheel, or rotor, that lies athwart two adjacent airstreams that have opposite direction. The rotor is massive and contains thousands of tiny passages parallel to the axis (and parallel to the airflows). Each sector of the rotor passes across one airstream and then the other; it accepts heat from one and gives up heat to the other. The heat-exchange surface is so large that the efficiency of sensible-heat recovery is 70 to 90%, typically. Because 80 or 90% of the frontal area of the rotor is open, the pneumatic resistance is low and low-power blowers may be used.

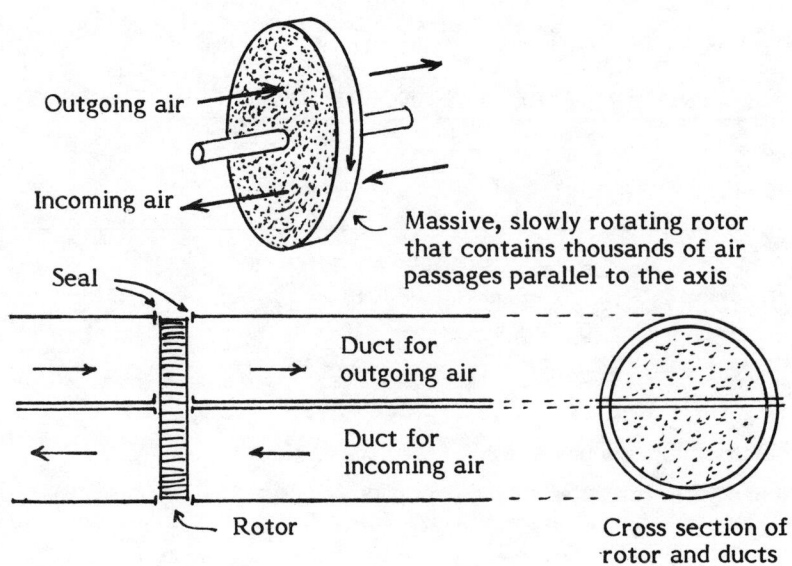

Outgoing air

Incoming air

Massive, slowly rotating rotor
that contains thousands of air
passages parallel to the axis

Seal

Duct for
outgoing air

Duct for
incoming air

Rotor

Cross section of
rotor and ducts

In some kinds of rotary exchangers, desiccant has been incorporated in the heat-transfer material or surfaces and accordingly water vapor may be absorbed or adsorbed from one airstream and delivered to the other. Thus latent heat, as well as sensible heat, may be recovered. The efficiency of overall recovery (enthalpy recovery) may be 70 to 90%.

ROTOR

The rotor of the household-type rotary exchanger (see Chap. 19) is 9 in. in diameter and 2.6 in. thick. Rotors for industrial exchangers may have diameters up to 14ft. and thicknesses up to 8 in.

The rotor may be of aluminum, plastic (e.g., a Teflon-base plastic such as DuPont's Nomex), or other material.

The rotor includes thousands of slender passages for airflow. The passages are parallel to the rotor axis. Their cross-sectional shapes may be random, or triangular, (honeycomb-like), or lenticular, as suggested by the following sketches.

Random   Triangular   Lenticular

Passage diameters may be about 1/24 in. for small exchangers and two or three times as great for larger ones. Total surface area of a 9-in.-dia. rotor with 1/24-in.-dia. passages is of the order of 100 ft^2. About 80% of the total face-area of the rotor may be open, for small rotors, and about 90% for large ones.

The rotor may be strengthened by bands and spokes of steel. Sturdy construction is needed so that the large temperature difference across the wheel will not cause it to warp and bind against the sealing strips (discussed below).

The heat-transfer surfaces may be impregnated with a hygroscopic material, or desiccant, such as LiCl or silica gel. The desiccant must be non-deliquescent, i.e., must not dissolve in water and float away. The desiccant picks up water from the (high-humidity) outgoing air and dispenses the water to the incoming air.

The rotor is rotated slowly -- at a few times to 20 times per minute -- by a small motor situated adjacent to it and connected to it by sheaves (pulleys) and a belt. In some installations the rotational speed is variable.

The typical linear speed of airflow in a passage in the rotor is of the order of 5 to 10 ft/sec.

DUCTS

The rotor is served by two parallel ducts, which may be semi-circular in cross section. The ducts carry the outgoing and incoming airstreams, which have opposite directions. Thus as the rotor turns, a given passage may find itself carrying outgoing air for a few seconds, then incoming air for a few seconds, and so on.

Each duct is served by its own centrifugal-type blower.

Plastic sealing strips are provided at each face of the rotor: they extend around the perimeter and also across the diameter that separates the two ducts. Such strip may consist of a compliant tube that presses gently against the rotor.

PURGE SECTOR

To reduce almost to zero the amount of air that can leak (at the rotor) from one airstream to another (thus posing the threat of contaminating the incoming air with toxic pollutants in the outgoing air), the designer sometimes provides a purge sector. This consists mainly of a baffle, shaped like a sector (piece of pie with 20° angle), mounted closely adjacent to the rotor at the location where the air passages progress from the outgoing airstream to the incoming airstream.

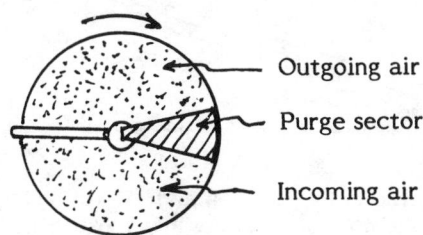

Outgoing air

Purge sector

Incoming air

82

FILTERS

In each airstream, a filter may be installed -- upstream from the rotor -- to minimize the amount of large-diameter dust particles that may strike, and adhere to, the rotor.

TEMPERATURE DISTRIBUTION ALONG A PASSAGE*

The temperature distribution along a typical passage in the rotor is complicated. The passage cross section is of complicated shape and very small area (of the order of 0.005 in.^{2}); the passage moves, every few seconds, out of one airstream and into the other, thus receiving warm outgoing air traveling in one direction and then cold incoming air traveling in the opposite direction. If the rotor contains desiccant, moisture is being taken up and given out by each portion of the rotor. All of these effects change when the speed of rotation changes, or the airflow speeds change. The goal of the designer is, of course, to provide an optimum combination of thermal mass, surface area, open area for airflow, rate of rotation of the rotor, rates of airflow in the associated ducts -- all in a compact space and at low manufacturing cost.

   I do not know of any detailed, clear, and reliable analysis of the temperature distributions in the passages of a typical rotor of a rotary exchanger, and do not myself know enough to specify the distributions. However, the basic principles of thermodynamics make clear some of the main features of the temperature distributions.

Case 1: Optimum Design and Operation Of A Rotor That Includes No Desiccant; No Condensation

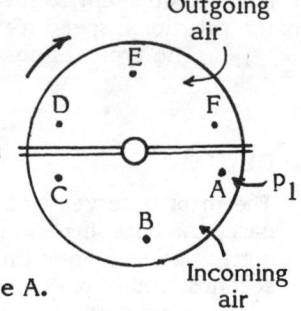

Consider a passage P_1 that has just moved into the incoming airstream. Assume, specifically, that it has reached location A indicated in the diagram, and some seconds later reaches locations B and C. Consider the air temperature at a given location along the passage. How does this temperature vary with the location along the passage -- assuming outdoor and indoor temperatures are 30°F and 70°F?

   Clearly, the variation, or distribution, must be somewhat as shown in Curve A of the following graph. Distance means distance (measured along the passage in question) from the cold (left) face of the rotor. The rotor is assumed to be 2 in. thick. At moments a few seconds later, indicated by locations B and C, the passage walls will be somewhat cooler and the air in the passage will be cooler also. Thus curves B and C are lower than Curve A.

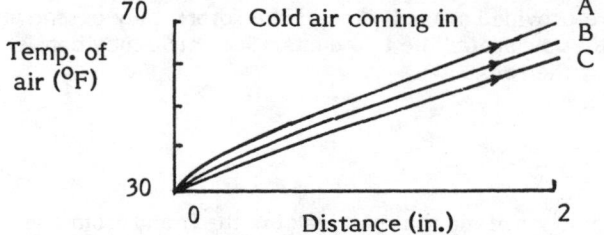

At a succession of moments after passage P_1 has moved into the stream of outgoing air, there is a corresponding set of temperature distribution curves, suggested by the following graph.

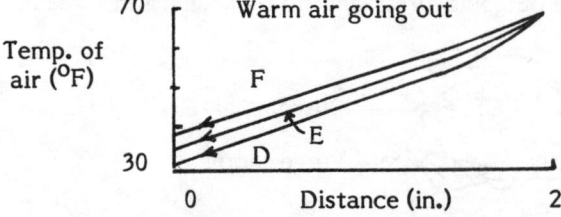

*Warning: Most of the ideas presented in this section and the following section consist of little more than surmise on my part. I have not come across any substantial reports on these topics.

Thus as the rotor rotates, the distributions of temperature along any one passage vary according to the succession of curves such as are shown in the two preceding graphs. Always the cold ends of the passages are nearly as cold as the outdoor air, and always the warm ends are nearly as warm as the room air; which implies that the efficiency of sensible heat transfer is high.

Case 2: Same, But Rotor Rotating Much Too Slowly

If the rotor were to rotate much too slowly, the curves would become more spread out vertically; in other words, the temperature changes at each location along any passage would be greater. The air emerging into the outdoors would not always be very close to outdoor temperature, and the air entering the room would not always be close to room temperature; in other words, the efficiency would be much lower.

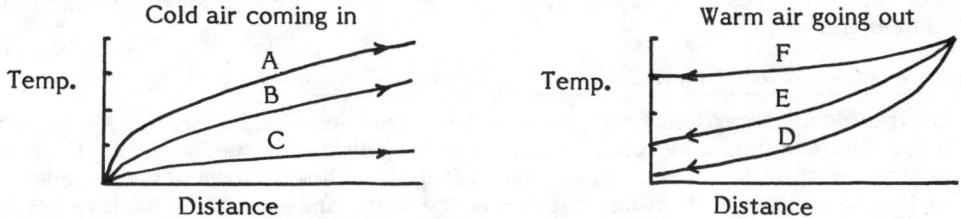

Case 3: Same, But Rotor Rotating Very Fast

In this case the airflows come in such short and rapidly alternating pulses that, at each location along a passage, the passage wall hardly has time to change temperature at all; the distribution of temperature-of-wall along the passage is practically a straight line: a straight line starting near 30°F at the cold face of the wheel and terminating at about 70°F at the warm face. Likewise the temperature distribution of the air in the passage corresponds approximately to a straight line, with similar slope. This implies, obviously, a very high efficiency of sensible-heat recovery. On each of the following graphs I have drawn only two lines, for simplicity.

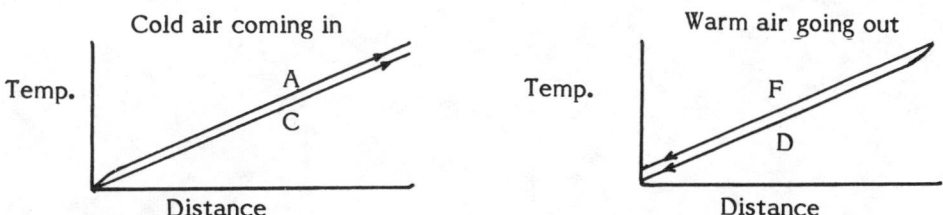

Note Concerning Symmetry

The remarks made in Chap. 10 concerning symmetry of the temperature distribution curves apply here also, i.e., even when latent heat as well as sensible heat is recovered. If no condensation or frosting occurs, then -- because the transfer process involved (transfer of sensible heat), is approximately linear -- the two sets of temperature distribution curves, i.e., for outgoing air and incoming air, must have symmetry; if you invert one set of curves and reverse it right for left, it must resemble the other set. Knowing this, one finds it easier to infer what the shapes of the curves must be.

84

CONSEQUENCES OF CHANGES IN ROTOR DESIGN AND OPERATION

What is the consequence of increasing the thermal mass of the rotor? More heat can be stored per half-cycle. Thus efficiency may be slightly increased and a lower speed of rotation may be acceptable.

What is the consequence of increasing the heat-transfer area? Heat intake and output are speeded up. Efficiency is increased and a faster speed of rotation may be acceptable.

What is the consequence of increasing the radius of the rotor, or increasing its thickness? Such changes increase the thermal mass and the heat-transfer area, with the consequences indicated in previous paragraph.

What is the consequence of increasing the rate of airflow through the rotor by increasing blower power? The rate of fresh air delivery to the rooms is increased, the amount of heat transferred is increased, and the efficiency is decreased. (The decrease in efficiency can be kept small if the rotation rate of the rotor is increased by the optimum amount; for example, if the rate of airflow is increased 20%, the speed of rotation should be increased by about 20%). The electrical power consumption by the blowers will be increased.

Note Concerning Consequences Of Fantastically High Rotational Speed

If the rotor were rotated at fantastically high speed, several troubles would arise. (1) Air entering a passage from one airstream might be sufficiently rotationally displaced that, when about to be discharged, it is confronted with the other airstream (oppositely traveling airstream) and is caught up in it -- joins it. Thus some stale air is brought back into the room, and some fresh air is redirected back to the outdoors. (2) The time intervals associated with reversing the flow directions in the passages -- instead of being relatively inconsequential -- become significant, representing a waste of time and blower power. (Trouble (1) could be avoided if the set of ducts on one side of the rotor were slightly rotated with respect to the set of ducts on the other side of the rotor. Trouble (2) cannot be avoided, except by reducing rotational speed.)

Chapter 13

AIRFLOW TECHNOLOGY

Introduction

Distribution of speed of laminar-flow air in a straight, circular-cross-section
 smooth-walled duct

Distribution for turbulent-flow air

Pressure drop in turbulent airflow in long straight circular-cross section duct

Effect of a 90-degree bend in duct

Pressure drop in turbulent airflow in long straight circular-cross-section duct

Centrifugal blowers: pressures and flowrates produced

Some laws of blower performance

INTRODUCTION

The designer of an air-to-air heat-exchanger wants the device to transfer much heat but to occupy
only a small space. Therefore he provides a large area of heat-transfer surface that is very com-
pactly arranged. He provides dozens or hundreds of slender passages in which the incoming air
and outgoing air flow, and he provides blowers to drive the two airflows.

The passages are usually so slender that, despite the fact that there are a great many of
them in parallel, there is a significant pressure drop -- of about 0.2 to 0.8 in. of water.

To bring incoming air to the entrance-ends of the appropriate set of passages a manifold
is needed. Another is needed to collect the air emerging from these passages. Likewise two manifolds are needed for the outgoing air. To design four manifolds that will not interfere with one
another, will entail few direction-changes of airflow (direction changes produce additional pressure drop), yet will occupy little space and be easy to fabricate, much ingenuity may be needed.

Air filters, situated upstream from the set of passages, are used to intercept any dust particles
that might clog the passages. The filters must be reasonably thin and open, otherwise they would
greatly increase the pressure drop.

Ducts, usually about 3 to 6 in. in diameter, may be needed to collect stale air from kitchen,
bathrooms etc. and deliver it to the exchanger. The ducts add to the overall pressure drop,
especially if they include many sharp bends.

In this chapter I discuss some general features of airflow in slender passages, ducts, etc.,
and I present data on blower performance. I start with a discussion of airflow in ducts, since this
is a well-understood subject

On the most important subject -- air flow in very slender passages -- I say very little. I
have encountered very little information on this subject.

DISTRIBUTION OF SPEED OF LAMINAR-FLOW AIR IN A STRAIGHT, CIRCULAR-CROSS-SECTION
SMOOTH-WALLED DUCT

When air is being driven steadily, in laminar flow, within a straight, circular-cross-section, smooth-
walled duct, the speed is greatest along the axis (centerline) and is approximately zero at the wall.
The speed varies "parabolically" with distance from the axis. See sketch.

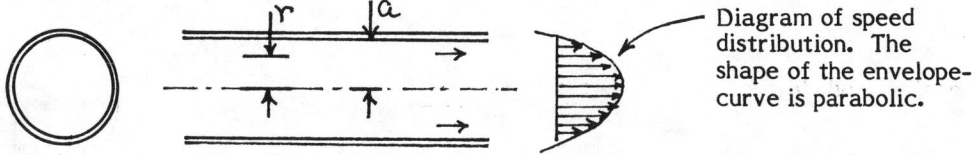

Diagram of speed
distribution. The
shape of the envelope-
curve is parabolic.

If the average speed of flow (over the cross section) is u_m and the duct radius is a, the speed u at some distance r from the axis is:

$$u = 2u_m(1 - r^2/a^2)$$

-- which, incidentally, implies that the speed of flow along the centerline of the duct is just twice the average speed.

Consider the quantity Q, which is the volume of air crossing a given cross section of the duct per second. Clearly: $Q = \pi a^2 u_m$. How does Q depend on the pressure head? Consider some long central segment of the duct: a segment of length L. Suppose the air pressures at the two ends of the segment are p_1 and p_2. Then it has been found that

$$Q = \pi a^2 u_m = \frac{\pi a^4 (p_1 - p_2)}{8 \mu L}$$

where μ is the viscosity of air. Notice that the volume flowrate is directly proportional to the pressure head, $(p_1 - p_2)$.

Warning: These equations do not hold near the ends of the duct. Also they do not hold when the flow is turbulent.

DISTRIBUTION FOR TURBULENT-FLOW AIR

Here the velocity-vs.-radial-position curve has a blunter peak, as suggested by the following sketch.

Here the speed-distribution curve is not parabolic.

The shape of the curve is not parabolic, and indeed has no simple algebraic description. Also, the average speed of the air (average taken over the cross section of the duct) is about 0.8 times -- not 0.5 times -- the speed of air traveling along the duct axis. My understanding is that the average speed (and likewise the volume flowrate) is proportional to the square root of the pressure head.

PRESSURE DROP IN TURBULENT AIRFLOW IN LONG STRAIGHT CIRCULAR-CROSS-SECTION DUCT

When air at sea level and 70°F is traveling (with turbulent flow) at a volume flowrate of 100 cfm within a 3-in.-diameter straight duct, there is a pressure drop along the duct (i.e., along the air-stream within the duct). The pressure drop per 100-ft.-length of duct is 3 in. of water (0.108 lb/in.^{2}). Incidentally, the linear velocity (averaged over the cross section of the duct) in this case is 34 ft/sec.

The following table shows the pressure drop (and linear airflow speed) for a variety of volume flowrates and duct diameters.

Pressure drop (in. of water) in turbulent airflow
in 100-ft-long, straight, circular-cross-section
duct. (Also linear speed of airflow (ft/min.))

Volume flowrate (ft³/min.)	8	16	32	64	128	256	512
Duct dia. 1½ in.							
Pressure drop	0.8	3	10				
Linear speed	(650)	(1300)	(2600)				
Duct dia. 2 in.							
	0.2	0.6	2.3	9			
	(400)	(700)	(1400)	(3000)			
Duct dia. 3 in.							
	0.025	0.1	0.3	1	4	13	
	(150)	(330)	(600)	(1200)	(2600)	(5000)	
Duct dia. 4 in.							
			0.09	0.3	1	3.5	
			(400)	(750)	(1500)	(3000)	
Duct dia. 5 in.							
				0.1	0.4	1.5	5
				(500)	(1000)	(1900)	(3800)
Duct dia. 6 in.							
					0.15	0.6	2
					(700)	(1400)	(2800)

(Source: S-45, p. 172)

Example What is the pressure drop in a 100-ft., 6-in.-dia., duct in which the volume rate of airflow is 512 ft³/min.? Answer: 2 inches of water, per last entry in table.

The pressure drop is approximately proportional to the length. Thus the pressure drop along a duct of any given length L can be found by multiplying the pertinent number from the table by (L/100 ft.)

Warning: The values listed above are for ducts that have extremely smooth walls. If the walls are rough, the pressure drops may be much greater.

EFFECT OF A 90-DEGREE BEND IN DUCT

Suppose that a circular-cross-section duct includes a 90-degree change in direction, accomplished by a smooth, quarter-circular segment the centerline of which has a radius of curvature that happens to be identical to the diameter of the duct. Experiments have shown that the added pressure drop (due to the 90-degree bend) is the same as that which would be produced by adding a straight section (of the same diameter) that has a length equal to 16 pipe-diameters. (It is assumed here that the linear speed of flow is great enough so that the flow is turbulent.)

Sometimes, 90-degree bends are accomplished by an elbow that has 2, 3, 4, or more straight segments, as suggested by the following sketches. Employing more segments (to make the bend a better approximation to a smooth quarter-circle) reduces the friction. Also, changing to an elbow of greater effective radius-of-curvature reduces the friction. Further reduction can be achieved by installing a set of "turning vanes".

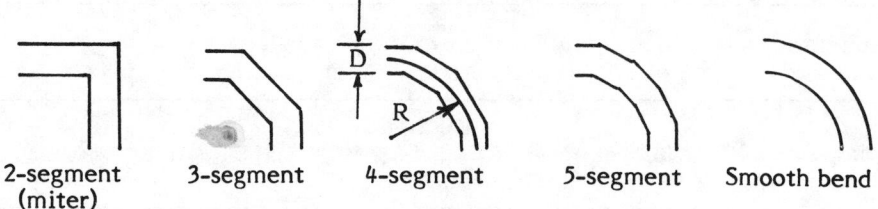

| 2-segment (miter) | 3-segment | 4-segment | 5-segment | Smooth bend |

Various types of elbows producing 90-degree bend
in a duct that is circular in cross section

The following table shows the pressure drop (in air flowing turbulently in a circular-cross-section duct) produced by 90-degree bend having 3, 4, or 5 segments in the elbow and having a variety of radius-of-curvature-of-centerline values. The ratio R/D is the ratio of the radius of curvature of the elbow centerline to the diameter of the duct.

Pressure drop in terms of: length, in pipe diameters, of
straight duct (of same diameter) that has same pressure drop

Ratio: R/D	1	1.5	2	3	4	5	6
3-segment elbow	21	17	16	17	18	18	19
4-segment elbow	19	14	12	11	10	10	10
5-segment elbow (or smooth elbow)	16	12	10	7	6	5	4

Source:
S-45, p. 174

Note: for 2-segment elbow (miter elbow) the pressure drop is 65 pipe diameters

Example What is the pressure drop in a 4-segment 90-deg. elbow in a 4-in.-diameter duct system if the radius of curvature of the elbow centerline is 16 in? Answer: Noting that the R/D ratio is 4 and referring to the 5'th column of the table, one sees that for a 4-segment elbow the value is 10 duct diameters, that is 40 in.

PRESSURE DROP IN TURBULENT AIRFLOW IN LONG STRAIGHT RECTANGULAR-CROSS-SECTION DUCT

To predict the pressure drop associated with turbulent airflow in a long straight duct that has rectangular cross section, one usually starts off by finding the equivalent circular-cross-section duct and then refers to the tables pertinent to circular-cross-section ducts. Experience shows, for example, that a duct 8 in. x 12 in. in cross section performs the same way a circular-cross-section duct 10.7 in. in diameter performs. Other correspondences are indicated below:

These (rect.) .... correspond to .... these (circ.) respectively

3 x 4,	3 x 5,	3 x 6,	3 x 8,	3 x 10	3.8,	4.2,	4.6,	5.2,	5.7	dia.
4 x 4,	4 x 5,	4 x 6,	4 x 8,	4 x 10	4.4,	4.9,	5.3,	6.1,	6.8	dia.
5 x 8,	5 x 10,	5 x 12,	5 x 14,	5 x 16	6.9,	7.6,	8.3,	8.9,	9.4	dia.
6 x 8,	6 x 10,	6 x 12,	6 x 14,	6 x 16	7.5,	8.4,	9.1,	9.8,	10.4	dia.
8 x 12,	8 x 14,	8 x 16,	8 x 18,	8 x 20	10.7,	11.5,	12.2,	12.9,	13.5	dia.
10 x 12,	10 x 14,	10 x 16,	10 x 18,	10 x 20	11.9,	12.9,	13.7,	14.5,	15.2	dia.

(Source: S-45, p. 178)

Example A duct that has a rectangular cross section 8 in. x 16 in. is equivalent, as far as pressure drop is concerned, to a circular-cross-section duct 12.2 in. in diameter. See underlined items in table.

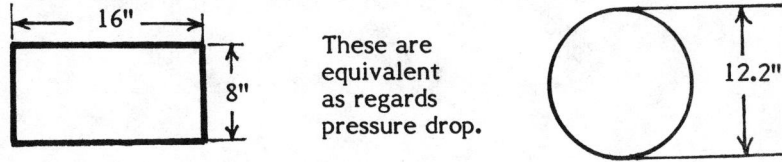

These are equivalent as regards pressure drop.

Formula Instead of using a table such as the above, one may use this formula:

$$d = (1.3 \text{ in.}) \left[\frac{\left(\dfrac{\ell \times w}{\text{in.}^2} \right)^{0.625}}{\left(\dfrac{\ell + w}{\text{in.}} \right)^{0.250}} \right]$$

where ℓ and w are the length and width of the rectangular cross section and d is the diameter of the equivalent circular-cross-section duct.

CENTRIFUGAL BLOWERS: PRESSURES AND FLOWRATES PRODUCED

The following table shows the approximate pressure heads and volume flowrates achieved by centrifugal-type blowers of various power ratings and various rotational speeds.

To provide a pressure	and flowrate of	use		
0.5 in. water	750 cfm	1/6 HP blower at	900	rpm
	1000	1/4	1000	
	1300	1/2	1200	
	1750	1	1500	
1.0 in. water	600 cfm	1/6 HP blower at	1050	rpm
	800	1/4	1100	
	1150	1/2	1300	
	1600	1	1550	
1.5 in. water	400 cfm	1/6 HP blower at	1300	rpm
	700	1/4	1300	
	1000	1/2	1400	
	1500	1	1600	
2.0 in. water	200 cfm	1/6 HP blower at	1500	rpm
	500	1/4	1500	
	900	1/2	1500	

Source: S-45, p. 180.

SOME LAWS OF BLOWER PERFORMANCE

Under some conditions of use of blowers that are producing turbulent airflow, these so-called laws are approximately valid:

To double the volume airflow, you must double the rotational speed of the blower, which quadruples the pressure head and requires eight times as much electrical power.

Or, in different words: volume airflow varies directly as the rotational speed, the pressure head varies as the square of the rotational speed, and the power supplied to the blower varies as the cube of the rotational speed.

Under some circumstances this law is not applicable.

For further details, see Bibl. item C-595 p. 10-9.

Chapter 14

HEAT-EXCHANGE TECHNOLOGY

Introduction

The two central problems

Heat-flow through sheet

Heat-flow through airfilms adjacent to sheet

INTRODUCTION

This chapter, concerned with the theory and technology of heat-exchange through walls that define very slender airspaces, is brief and of little use. It is brief for several reasons, the most important of which is that I know very little about the subject.

Another good reason is that the subject is so complicated. It is complicated because:

The airspaces involved are so slender. Most textbooks on heatflow and heat-exchange deal with generously wide passages for airflow. When the passages are very slender, surface roughness becomes relatively more important. Yet roughness is hard to define, hard to measure, hard to deal with in engineering equations.

In many exchangers that have very slender passages, the passage cross-sections are not round: they may be rectangular, triangular, or lens-shaped, or may have other shape to which standard formulas do not apply.

The slender passages are served by large manifolds which in turn have peculiar shapes -- hard to deal with in engineering equations.

Where a duct joins a manifold, and also where a manifold joins a set of slender passages, the behavior of flowing air is hard to predict, hard to measure, hard to describe.

The patterns of airflow may vary if the speed of airflow is varied.

In some of the most cost-effective exchangers the passages for airflow are only 1/16 or 1/24 in. in diameter, and accordingly the boundary airfilms, or stagnant outer portions of each slender airstream, may have thicknesses almost comparable to the passage diameter. Thus the task of understanding in detail how the air flows and the heat flows is made especially difficult.

If frost forms in the passages, the patterns of airflow and heatflow change -- change more or less unpredictably.

If the exchanger is of a type that recovers water, the situation is even more complicated.

My guess is that a person designing a small exchanger, i.e., one that has very slender passages, would do well to try to find just how well the existing slender-passage exchangers perform, then try to find how to slightly vary the design so as to achieve even better performance. That is, he should proceed empirically in the light of information obtained with various existing exchangers. My impression is that, in the last analysis, most of the classical formulas for heatflow through small-passage sheet-and-airfilm systems are empirical; accordingly, relying on actual experimental data on existing exchangers may be fully as rewarding as relying on formulas that may not be entirely applicable.

Some detailed analyses, employing formulas, of exchanger performance have been made by The Memphremagog Group. Perhaps their reports would be of interest to other designers.

The following sections present various bits of information that may be pertinent.

THE TWO CENTRAL PROBLEMS

In designing an air-to-air heat-exchanger, an engineer must try to predict:

1) How much heat will flow through a typical, thin, heat-transfer sheet and its adjacent airfilms when the temperature difference has a given value and the two flowrates have a given value.

2) What the temperature difference will be at each location along the exchanger when the two flowrates have a given value and the indoor and outdoor temperature differences at various locations are not known a priori.

HEAT-FLOW THROUGH SHEET

Heat-Flow Through A Sheet Of Metal

The amount of heat conducted through a sheet of metal is proportional to the thermal conductivity k, and to the area through which the heat flows, and to the difference in temperature of the <u>two faces</u>, and to the thinness (i.e., 1/thickness), and to the duration of flow. That is,

$$Q = k \ A \ \Delta T \ (1/x)(\text{duration})$$

The time-rate of flow is:

$$q = k \ A \ \Delta T \ (1/x)$$

<u>Example</u> A sheet of copper has an area of 3 sq. ft. and a thickness of 0.01 in. The two faces are maintained at 70°F and 72°F. The conductivity of the copper is 2500 (Btu in.)/(ft^{2}hr °F). What is the rate of heat-flow through the sheet? Since ΔT is 2 F deg. and 1/x is 1/0.01 = 100, the answer is:

$$q = (2500)(3)(2)(100) = 1,500,000 \ \text{Btu/hr}.$$

Two Warnings

(1) The equation presented is valid only when one knows the actual temperatures of the actual faces of the sheet in question. Merely to know the temperatures of the adjacent airstreams -- at, say, a distance of 1/16 in. from the sheet -- is not sufficient. Obviously, it is often difficult or impossible to find the temperatures of the actual faces of the sheet, and the equation is then of dubious usefulness. (2) The equation pertains to conductivity heatflow only, and accordingly if the sheet is somewhat transparent to radiation such as is emitted by all objects that are at about room temperature (3 to 30 micron radiation), the total heatflow may be much larger than the equation would suggest.

Heat-Flow Through Sheet Of Other Material

The equation presented above applies not only to sheets of metal but also to sheets of other materials, e.g., plastics. One big difference is that plastics have about a thousand-fold lower conductivity than aluminum or copper has. Conductivities of various pertinent materials are listed in the following table.

Approximate Value of Conductivity k and Resistivity (1/k) of Some Common Materials

Material	k $\left(\dfrac{\text{Btu in.}}{\text{ft}^2 \text{ hr }^\circ\text{F}}\right)$	$1/k$ $\left(\dfrac{\text{ft}^2\text{hr }^\circ\text{F}}{\text{Btu in.}}\right)$
Aluminum	1400	0.0007
Copper	2500	0.0004
Steel	400	0.0025
Glass	10	0.1
Brick	5 to 10	0.1 to 0.2
Lucite or Plexiglas	2	0.5
Fiberglass	0.3	3
Oak wood	1.1	0.9
Air (with no convection, no radiation)		
at 32°F	0.14	7
at 212°F	0.18	5.5

Heat-Flow Through The Wall Of A Tube

Heat-flow through a thin sheet of metal is practically the same whether the sheet is flat or is bent around so as to form a circular-cross-section tube (pipe or duct), if the temperature difference across the sheet is the same in each case.

Accordingly the following formula, derived for a flat sheet, applies also to a tube:

$$q = k \ A \ \Delta T \ (1/x).$$

A small question arises concerning A, the area of the tube wall. How is this area to be computed? Specifically, how is the radius of the tube to be computed? Of course, if the tube wall is very thin, one gets almost the same answer irrespective of whether, in computing area, one uses the inner radius, outside radius, or the average of these. Often one uses the average radius, and the area A then is:

$$A = 2\pi \ (\text{ave. radius of tube})(\text{length of tube})$$

However, if the tube wall is very thick (outer radius at least twice as great as the inner radius), one should use -- not the simple average radius -- but the logarithmic radius. See **Ref. M-86**, p. 12.

HEAT-FLOW THROUGH AIRFILMS ADJACENT TO SHEET

Here matters become especially complicated. They are over my head. See the book by Chapman (C-205).

Chapter 15

AIR DISTRIBUTION STRATEGIES

Introduction

Location of exchanger

Location of outdoor ends of intake and exhaust ducts

Indoor distribution strategies

Where, exactly, is fresh air most needed?

Stopping the pollution at its source

Insulation, screens, and dampers for ducts running to outdoors

Air-change goals and replacivity goals

Application of exchanger to NCAT 800-ft^2 Micro-Load House

INTRODUCTION

Delivering fresh air to the house is only part of the task. Another part is to deliver it exactly where it is needed, when needed.

LOCATION OF EXCHANGER

A very small exchanger may be mounted directly on an outside wall - of living room, bathroom, or other room. See Chapters 19 and 20 for examples.

Larger exchangers are usually installed just below the floor or above the ceiling or in a utility room. The exchanger should be located close to an outside wall, so that the fresh-air intake duct can be short. If it is short (and of course, well insulated), the fresh air entering the exchanger will be nearly as cold as outdoor air and accordingly will be very effective at extracting heat from the outgoing air.

Before choosing the location, the architect or homeowner should consider carefully just where fresh air is most needed.

LOCATIONS OF OUTDOOR ENDS OF INTAKE AND EXHAUST DUCTS

These may be only a few feet apart, or may be 10 or 20 feet apart, as suggested by the following diagrams. The farther apart, the smaller the chance that some of the exhaust air will find its way back to the intake duct. If there is some nearby, outdoor, source of pollution, the intake duct should be situated as far from this source as possible.

If the outdoor ends of the ducts are equipped with downward-turned terminal elbows or suitable hoods, rain and snow will be excluded.

96

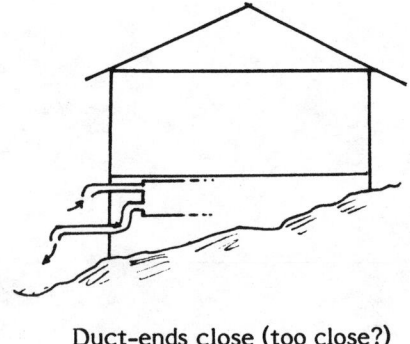

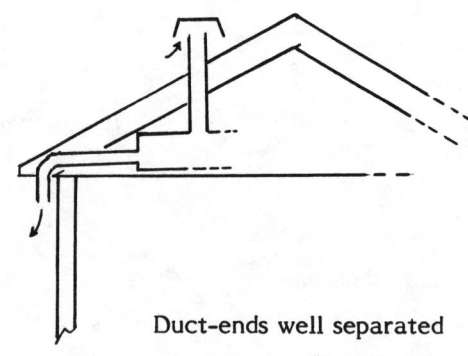

Duct-ends close (too close?)
together

Duct-ends well separated

It may be advantageous to have the intake duct on an upwind side of the house and have the exhaust duct on a downwind side. There may be advantage in having the intake duct on the south side of the house, so that the air that is taken in will be somewhat warmer than air on, say, the north side.

INDOOR DISTRIBUTION STRATEGIES

No-Duct Distribution

Many exchangers are of small capacity and employ no ducts. Fresh air from the exchanger flows directly into the room; usually this air emerges from the lower part of the exchanger and spreads outward and somewhat downward from the exchanger; being somewhat colder than room air, the incoming air tends to descend toward the floor -- accordingly the room occupants feel no current of cold air past their faces. Stale air is taken in by the exchanger via a grill in the upper part of the exchanger.

How much of the fresh air will reach kitchen, bathrooms, bedrooms? The answer may be much or little depending on how far away these rooms are and whether there are intervening closed doors.

How is the stale air collected by the exchanger? Collection is passive -- and uncertain and delayed. Whatever pollutants diffuse to the neighborhood of the exchanger may be expelled to outdoors; other quantities of pollutants may remain in the house for long periods.

Some users purchase two exchangers and install them in two different rooms: the living room and one other room, say the master's bedroom.

Distribution Using Ducts

Often ducts are used, especially if the exchanger's capacity is large -- larger than is needed for just a living room.

There is this difficult choice: Should the duct system serve the fresh air, i.e., serve the delivery of air, or should it serve the stale air, i.e., serve the collection and expulsion of air? Or should there be two duct systems, one for fresh air and one for stale air?

Ducts used to deliver fresh air: Usually, the duct system is employed for delivery of fresh air if the house already has a duct system, e.g., for blower-forced delivery of hot air from a hot-air furnace or for blower-forced delivery (in summer) of cool air from a central air-cooler. The reasoning is simple: "The duct system is already here and is already delivering air to the rooms; let's put it to the additional use of distributing fresh air from the exchanger." When the hot-air-furnace blower is operating, it insures quick delivery of fresh air to many rooms; when the blower is not running, the delivery of fresh air continues because of the continual operation of the blowers within the exchanger.

The fresh air from the exchanger is normally introduced (to the furnace duct system) just upstream from the furnace; thus the fresh air is heated by passage through the furnace plenums and, before entering the rooms, is as hot as may be desired.

Some drawbacks to this system are: (1) If airflow to some rooms has been shut off (by dampers), those rooms will receive no fresh air; even although a room may not need heat, it may need fresh air. (2) The powerful hot-air-furnace blower may create imbalance of the two air-flows in the exchanger: that blower tends to speed up the flow of incoming air but does not speed up the flow of outgoing air. (3) Elimination of polluted air from kitchen, bathroom, etc., is uncertain and delayed.

Ducts used to collect stale air: This strategy has the great merit that the most highly polluted air (for example, in kitchen and bathrooms) is collected immediately, locally. It is not given a chance to (or invited and encouraged to) diffuse into the other rooms prior to being collected and expelled. Fresh air may be delivered to one location (in living room, say) and permitted to diffuse into the other rooms. (Flow to other rooms may be negligible if the doors to those rooms are shut.) The ducts that carry the stale air to the exchanger do not need to be insulated.

Use of two duct systems, one to collect stale air and the other to deliver fresh air, can sometimes be worthwhile. Usually, however, one duct system is enough.

WHERE, EXACTLY, IS FRESH AIR MOST NEEDED?

The answer depends on the location of the house (the kind of earth it rests on, the direction of the prevailing wind), the size and shape of the house, the materials used in construction, the kinds of uses to which the rooms are put and the locations of the rooms, the extent to which the rooms are isolated from one another by walls and closed doors, the types of stove and furnace, the habits of the occupants (number of long showers taken, number of persons smoking, frequency with which cooking of cabbages, broccoli, and other oderiferous forms of food occurs, any hobbies that involve release of pollutants), and any special health problems (especially pulmonary problems) the occupants have.

Some Pertinent Questions

Kitchen	Gas Stove?
	Much used?
	Much cooking of smelly vegetable?
	Does kitchen have its own air-exhaust system?
	Is this put to use frequently?
	Window often left partly open?
Bathroom	Much used?
	Long showers?
	Does bathroom have its own air-exhaust system?
	Is this put to use frequently?
	Window often left partly open?
Living room	Much used?
	Much smoking?
	Many sofas, chairs, rugs, etc., that contain formaldehyde?
Bedrooms	Many materials that contain formaldehyde?
	Windows left open all night?

Basement or crawl- space	Damp? Gas furnace? Oil furnace? Leaky oil tank? Does earth beneath basement contain much radon-emitting material? Is the terrain granitic? Does basement have a floor that is impervious to radon?

Is house very airtight?

Are vapor barriers used? Just in outer walls? In ceiling also?
  Between first-story floor and basement? Between basement and earth
  beneath it?

Has urea formaldehyde insulation been used?

Is the house in a windy location?

STOPPING THE POLLUTION AT ITS SOURCE

It is sometimes feasible (and economical) to "stop the pollution at its source." If pollutants originate just at the cooking stove, or just in a bathroom, one can vent just these areas (by use of small exhaust fans, or by opening windows briefly) and leave the house as a whole alone: no air-to-air heat-exchanger is needed. Likewise if the only source of pollutants is the crawlspace (say much radon or much moisture is present here and threatens to flow upward into the rooms), one can provide large vents for the crawlspace; or, if there is no insulation between crawlspace and the rooms, install an exchanger that will serve just the crawlspace; then the living region of the house may not need an exchanger.

Usually there are many sources of pollutants, and to stop them all at their sources would not be practical. The problem is solved by a use of an exchanger that serves the house as a whole.

INSULATION, SCREENS, AND DAMPERS FOR DUCTS RUNNING TO OUTDOORS

The ducts running from the exchanger to the outdoors should be insulated. The cold-air intake duct and its insulation should be covered by a vapor barrier and this should be well sealed.

Both of the outdoor duct-ends should be equipped with screens that will exclude animals, flies, bees, etc.

If the exchanger is to be left off for long periods in winter, backdraft dampers should be installed. Otherwise, cold air can find its way into the rooms via the (inoperative) exchanger.

AIR CHANGE GOALS AND REPLACIVITY GOALS

Past Goals

Ten or twenty years ago the goals pertinent to indoor air in houses were often expressed in terms of air input per hour. (Obviously this is unfortunate; what counts is replacivity, and high rate of air input may or may not correlate with high replacivity.) Typical goals were: 2, or 1, or ½ house-volumes per hour.

Sometimes the goals were expressed in terms of air input per hour per occupant. This makes good sense if the pollutants are generated by the occupants or by occupant-related activities. It makes no sense if the pollutants come from the earth or concrete basement, or Trombe wall, or furnishings containing formaldehyde.

Today's And Tomorrow's Goals

These should be chosen carefully and logically -- on the basis of demonstrated need with respect to health and comfort and the more obvious need to reduce waste of heat. Also, the goals should be expressed, first, in terms of replacity, and second (where the correlation between replacity and rate of air input can be estimated with reasonable accuracy) in terms of rate of air input.

One good procedure would be for the house occupants to try out some arbitrary choice of replacity and then ask a technician to evaluate the actual quality of the indoor air. The occupants can start with some very low replacity and then increase it until the technician finds the air quality to be satisfactory. Of course, a needlessly high replacity should be avoided: it entails waste of heat, waste of money.

My guess is that a replacity of 40% will be about right (i.e., the best compromise) for most houses, 15% may be about right for houses in which most of the circumstances are favorable, and 80% may be about right for houses in which circumstances are moderately unfavorable. Of course, one's opinion as to what is the best compromise will change as the cost of auxiliary heat changes and as the occupants' wealth changes.

I expect that, for many decades to come, views as to what constitutes the best compromise will change continually as medical men learn more about people's reactions to the various pollutants and as new chemicals come into use in house furnishings and building materials.

APPLICATION OF EXCHANGER TO NCAT 800 FT^2 MICRO-LOAD HOUSE

The National Center for Appropriate Technology issued, in mid-1981, an 11-page set of 3 ft. x 2 ft. plans of a proposed superinsulated house: a very tight, one-story house. The design calls for use of a Des Champs 79M4-RU air-to-air heat-exchanger which is suspended beneath the floor.

The incoming fresh air, after being driven through the exchanger, is delivered to a plenum extending above the ceiling of the central utility room and the ceiling of a small central hallway serving the two bedrooms. The plenum, defined by a main ceiling and a drop-ceiling, is 6 in. in inside height. Air flows from this plenum, via four 12 in. x 2¼ in. grills, to the living-dining area, the two bedrooms, and the bathroom.

Stale air from the bathroom enters (via a small grill 5 ft. above floor level) a 4-in.-diameter flexible duct, which later widens to become a 6-in.-diameter duct. Stale air from the living-dining area enters this duct via a larger grill. A one-inch gap is left at the bottom of the bedroom doors to allow stale air to escape therefrom into the living-dining area. (One wonders whether such gaps provide problems, such as transmitting noise, light, odors.)

The exchanger runs whenever the humidistat senses excessive humidity and whenever the bathroom door is shut.

The stale air is vented to outdoors via a roof stack: a vertical, 6-in.-diameter stack situated 2 ft. north of the ridge of the roof.

The fresh air intake is in the upper part of the east wall of the foundation.

Chapter 16

SENSORS, CONTROLS, AND FAIL-SAFE DEVICES

INTRODUCTION

How is the rate of supply of fresh air to be controlled? Manually? Or automatically? If automatic control is used, what parameter should it be based on and what kind of sensor should be used? What happens if the device fails, or the electric power fails? Is some kind of failure-indicator needed? Is some kind of automatic correction of failure, or crude compensation for failure, needed?

Such questions are addressed in this chapter.

MANUAL CONTROL

Obviously the exchanger can be controlled manually: turned on or off, or changed from low speed to high speed, by a member of the household. This approach is simple and inexpensive, but if the members of the household are careless or forgetful, performance may be poor: not enough fresh air sometimes, too much (with consequent waste of heat) at other times. Also, the need to keep the exchanger in mind and the need to make decisions as to whether to change the switch setting may prove to be a burden.

Besides changing the speed of the exchanger, the members of the household may adjust dampers so as to redistribute the fresh air (or removal of stale air) in different parts of the house at different times of day. Focus on the living room during the day and the bedrooms at night, for example.

CONTROL BY TIMER

A device that contains an electric clock may be employed to change the exchanger speed at different times of day. It could control, also, the dampers that govern the distribution of air to (or from) the rooms. (Timers are already being used to control the distribution of cold air to rooms, in the summer; see the article "Programmed Cooling" in the July, 1981, Popular Science, p. 24.)

CONTROL BY TEMPERATURE

Using an outdoor thermometer, one might arrange to reduce blower-power whenever the outdoor temperature is below, say, 30°F. This would save some money, inasmuch as the exchanger's efficiency is well below 100% and thus significant loss of heat occurs whenever the outdoor temperature is very low. It makes good sense to be generous with fresh air on warm days and stingy on very cold days. Also, reducing the speed of exchanger operation on very cold days may be permissible inasmuch as infiltration of natural type is then at a maximum, thanks to chimney effect.

Some exchanger manufacturers supply a temperature sensor that is used to sense the threat of air-blockage by frost. When there is partial blockage of the outgoing-air passages, the incoming air entering the room is colder than it normally is; a thermometer that detects this situation can take constructive action -- such as slowing or stopping the blower driving the incoming air.

CONTROL BY HUMIDITY

A humidistat (or dehumidistat) may be used to turn on an exchanger, or change the speed from low to high, whenever the relative humidity of room air exceeds a certain value -- say 50%. (This remark applies, obviously, to exchangers that do not recover water. It applies also, but to a lesser extent, to exchangers that do recover water -- inasmuch as the recovery is far less than 100%).

Several exchanger manufacturers recommend use of a dehumidistat.

CONTROL BY CONCENTRATION OF POLLUTANT

Ideally, the rooms would include a set of sensors, each responsive to a different major pollutant, and each capable of turning on the exchanger (or increasing its speed of operation) whenever a pollutant-concentration is found to exceed a specified limit. This approach might be practical with respect to a few pollutants -- in certain special situations existing today: e.g., a large and important building containing many people, with known serious pollution-threat. Perhaps the practicality will be widespread in another 10 or 20 years.

CONTROL BY UTILITY ACTIVATION

Arrangements could be made wherby the exchanger is turned on, or turned up, whenever a bathroom door is closed, or a cooking stove is activated, or a clothes dryer is activated, or electric power is turned on in a workshop or hobby room.

INVERSE-OF-WINDSPEED CONTROL

There is obvious merit in the idea of having the speed of the exchanger blowers diminish as the outdoor windspeed increases. Every house has some natural infiltration, and this increases with windspeed. What is important, in keeping levels of pollution low, is the total rate of fresh-air input: the greater the natural input, the smaller the input needed from the exchanger. (If the house occupant ignores this, he will be wasting electrical power and heat whenever the windspeed is great.)

It would not be difficult to employ an above-roof-mounted hot-wire anemometer and arrange for the output signal to reduce blower power as the windspeed increases. At some very high speed, say 25 mph, of the wind, the exchanger should be shut off entirely, in certain not-extremely-tight houses.

Note: In a report dated 9/14/81 I described a passive type of "inverse vent": a house vent which would automatically and passively gradually close as windspeed increases, the extent-of-closure-as-a-function-of-windspeed being such that the overall rate of air-change (via cracks, etc., and via the special vent) would remain constant, irrespective of windspeed.

Clearly, such a device would slightly reduce the need for an air-to-air heat-exchanger.

CONTROL FOR ENCOURAGING CONDENSATION OF MOISTURE BUT PREVENTING FROST ACCUMULATION

Perhaps it would be feasible to provide a control that would regulate airflow rates (or regulate preheating or other parameter) in such a way that, under typical conditions in winter, much condensation of moisture would occur but no frost would accumulate. The process of condensation may be welcomed if it leads to some recovery of latent heat, but the process of frost formation may be most unwelcome inasmuch as it may lead to clogging of certain passages. The line between condensate formation and frost formation may be a narrow one, yet it might be feasible to design a control that would encourage operation that comes close to the line but does not cross it!

DEVICES FOR WARNING OF POWER LOSS OR EXCHANGER FAILURE

Suppose that the exchanger routinely used in a very tight house fails: a blower motor fails, the electric supply fails, the air passages become completely blocked, or some other misadventure befalls.

Will the house-occupants remain unaware of such failure? Will the concentrations of pollutants build up indefinitely? Will the health of the occupants be impaired?

One may expect that, ordinarily, no harm will result. The pollutant concentrations will not be so very excessive; the occupants may notice the build-up of smells or humidity; they may notice that the noise of the blowers has ceased; they may notice that the entire house is without electric power. Thus they may take suitable steps, such as opening a window or quitting the house.

But the designer must expect that failure of the exchanger may indeed occur. He may feel obliged to provide some equipment that, at the least, will warn the occupants that there is trouble.

Such trouble could be sensed by a device that detects a drop in voltage or detects a drop in air pressure close downstream from a blower.

Visual Devices

Whenever the exchanger is operating a green light may be lit, to show that operation is proceeding smoothly. If the light goes out, the occupants should take action. (But will they notice that the light goes out? Perhaps they are in another room, or are asleep.)

Alternatively, a very bright light might be turned on automatically whenever the general electric supply fails. The bright-light bulb would be energized by a battery. (But might not the battery be dead when the time comes for it to do its duty? And might not the bright light be unnoticed, if the occupants are asleep or in a different room?)

A manometer connected so as to indicate the pressure drop across the heat exchanger proper would give helpful information, e.g., information as to blower failure or clogging of the outgoing-air passages by dust or frost.

Ribbons hung in the fresh-air stream emerging from the exchanger might do nearly as good a job. If the ribbons stop moving, no fresh air is coming in!

Audible Devices

Battery-operated buzzers, bells, or the like might be used to indicate loss of power. However, such system may be ineffective if the occupants are deaf, or are in a remote room, or if the batteries are dead.

FAIL-SAFE ACCESSORY

It occurred to me early in 1981 that a truly fail-safe accessory for air-to-air heat-exchangers could be provided.

My idea was that if, for any reason, the exchanger ceases to operate, a small port (or door, or cover) in the outside wall of the house would open, allowing direct natural inflow of fresh air.

The port might be 8 in. in diameter. A continuous torque urging the port to open could be exerted by gravity, or by a spring or other means. This torque would be opposed by the "velocity pressure" of one of the two airstreams. As long as the airstream exists, the port stays closed. But if the airstream stops for any reason, the port opens.

Velocity pressure alone would probably be too slight to open the port -- but strong enough to trigger some far more forceful actuator, e.g., a spring-type actuator that is manually re-readied after each use.

Instead of a single port, two ports might be provided -- one on the west side of the house and one on the east. Thus when the exchanger fails and the ports open, a through-draft of fresh air can exist. The prevailing wind can speed the flow. Alternatively, one port might be high up and one low down, to take advantage of chimney effect.

Chapter 17

COMPARISON OF ROTARY AND FIXED-TYPE EXCHANGERS

INTRODUCTION

Which kind of exchanger is better: one employing a rotor or one employing a fixed set of heat-transfer surfaces? This is one of the first questions that anyone considering buying an exchanger should ask.

Unfortunately there is no simple answer. Each type of exchanger has strong points and weak points. Several types of exchangers are so new that there is little reliable information on the different aspects of performance.

At this stage, one cannot do much more than point out some of the main questions, or issues. Such questions are listed below.

COMPACT, LAMINAR-FLOW EXCHANGER WITH FIXED TRANSFER SURFACES: SOME QUESTIONS

Such devices may employ very slender passages, i.e., passages only about 1/16 in. in effective diameter. The passages must be slender if the flow is to be at highspeed and yet is to be laminar. Because the passages are so slender, the assembly must include hundreds of tiny partitions or spacers to insure that the (very small) spacings be maintained, i.e., that sheets meant to be about 1/16-in. apart do not warp and touch one another, blocking the airflow. Usually, to form the passages and provide the partitions, a deeply and finely corrugated sheet is employed: it is interposed between two simple flat sheets.

It is not feasible to have air travel in opposite directions in adjacent passages defined by a corrugated sheet: the manifolds would be too complicated, inasmuch as there are several thousand such passages. Consequently the usual practice is to have the same direction of air, and same class of air (for example, stale air), in all of the passages defined by a given corrugated sheet.

The exchanger can be made especially compact and small if the two airstreams are in cross-flow orientation, rather than counterflow orientation.

With respect to such exchangers, these questions may be asked:

1. Is heat-exchange efficiency significantly reduced by the use of cross-flow rather than counterflow?

2. Is heat-exchange efficiency significantly reduced by the fact that about 60% of the wall area is area of the corrugated sheet, i.e., a sheet not lying immediately between the two classes of air? The corrugated-sheet surface provides friction, and it also provides some useful conduction, i.e., acts somewhat as a fin. Is there a significant penalty associated with the use of (corrugated) sheets both sides of which are bathed by the same class of air?

3. Is there any significant risk that there will be (initially, or after a few years of use) some small holes or cracks through the flat sheets, with the consequence that stale air can migrate into the fresh air?

4. Is there a significant tendency for dust particles to lodge permanently on the heat-transfer sheets -- in view of the fact that the direction of airflow never changes?

5. In exchangers in which the sheets are permeable to water (with the consequence -- desirable in houses that tend to be too dry -- that water can be transferred from one airstream to the other), the exchange of water is reduced by the fact that a large fraction of the surface area is area of the corrugated sheets. Because

the surfaces of such sheet are in contact with the same class of air, the sheets contribute very little to the transfer of water. Is this a serious handicap? Or is an ample amount of water transferred through the flat sheets?

6. If the outdoor air is very cold, frost may form within the cold portions of the passages for outgoing air, thus restricting airflow here. (The house occupants may not know that this is happening.) Will this cause a rapid increase in frost formation? That is, will the reduction in flow of outgoing air mean a reduction in warming of the incoming air -- with the result that the cooling of the outgoing air will be accentuated and frost formation will be accelerated? Is there, here, a kind of instability, associated with the fact that clogging of the outgoing-air passages is <u>not</u> compensated by a clogging of the incoming-air passages? (There are various routine schemes for detecting and discouraging or preventing frost formation. But how much do they add to the cost and complexity?)

7. In an exchanger in which the sheets are permeable to water, to what extent do pollutants dissolve in the water and pass (with it) from the stale air to the fresh air? To what extent do pollutants pass from stale air to fresh air (via these sheets) even when no transfer of water is occurring?

Notice, by the way, that the <u>rotary</u> exchanger avoids many of these limitations or questions:

Counterflow -- not crossflow -- is used.

All of the sheets, including the corrugated ones, contribute fully to heat-transfer.

Holes or cracks in the sheets do not lead to any cross contamination.

All of the surfaces may be desiccant impregnated. Thus all may contribute to water transfer.

In any given passage, the direction of airflow reverses every few seconds -- which may (?) tend to help release adhering dust.

Such reversal may (?) tend to prevent accumulation of liquid condensate and frost.

When and if some clogging of the passages does occur, as a result of frost accumulation, it occurs equally with respect to outgoing air and incoming air (inasmuch as the two kinds of air use the identical passages -- sequentially). Thus there is, I expect, little or no acceleration of frost formation. Almost never, I expect, will a passage become entirely closed off or will frost be "locked in."

Rotary exchangers have their own limitations, and one may ask questions such as these:

1. Is it difficult to make the faces of the rotor sufficiently smooth and accurately perpendicular to the axle -- so that the pertinent seals can perform well?

2. Is it easy to make seals that, even after years of use (continually rubbing against the rotating rotor), will remain effective? Will there be appreciable wear or embrittlement or warping?

3. Are there any circumstances where the wiping action of the sealing strips will increase any tendency for dust etc. to adhere to, and partially clog, the passage-ends?

4. Is the rotor subject to stalling, for example when frost accumulates?

I have no knowledge of any actual instances where troubles of these kinds have arisen.

QUESTIONS CONCERNING NON-COMPACT EXCHANGERS OF FIXED TYPE

Exchangers that are of fixed type and employ airspaces of more generous minimum dimension (airspaces that are, say, 1/6 to 1/2 in. thick) can easily be designed so as to employ counterflow. Interleaved corrugated sheets are not needed. Dust accumulation may be almost negligible.

These questions may be asked:

1. Do such exchangers tend to be considerably more bulky than the exchangers discussed in the previous section? Are they too bulky for mounting directly in a living room, kitchen, bedroom, or bathroom? Are they also significantly heavier, harder to handle, harder to ship?

2. Do they almost invariably employ heat-transfer sheets (of aluminum or plastic) that are impermeable to water? That is, is the recovery of latent heat ruled out -- except when actual condensation of water occurs? Is the recovery of water ruled out?

My guess is that rotary-type exchangers can, in principle, be a little more compact and a little easier to handle and are a little more suitable for in-room mounting. Certainly they are highly suited to recovery of latent heat and recovery of water.

Note, however, that it is unfair to criticize a fixed-type exchanger because of its bulk if the bulk is associated with considerably higher flowrates such as may effectively serve an entire house rather than, say, just a living room.

***

In attempting to compare various types of exchangers, one should consider many other matters also. See long list of topics presented in Chapter 32.

Chapter 18

GENERAL COMMERCIAL AVAILABILITY OF EXCHANGERS

Although the production and sale of large-size air-to-air heat-exchangers for use in industrial plants, office buildings, hospitals, etc., has been an important activity for many decades, small exchangers -- small enough and economical enough for use in houses -- are relatively new.

I have assigned one chapter to each company that (a) produces exchangers suitable for use in houses and (b) has provided me with detailed information on its products.

In Chapter 28 I have assigned individual sections to companies that (a) make very large exchangers for industrial use etc. but do not make small exchangers for use in houses, or (b) are still wholly engrossed in design and development work on exchangers, or (c) have failed to supply me with detailed information.

The figures presented, for any given exchanger, for the airflow, efficiencies, etc., are figures supplied by the manufacturer. To what extent, if any, such figures are over-optimistic, I do not know.

I have tried to include prices. However, prices have a way of increasing whenever one turns one's back.

I have tried to show the internal workings of the exchangers. Certainly my drawings are incomplete and inaccurate. Mainly I have focused on explaining the general principle of design, principle of operation.

Nearly all of the exchangers described employ counterflow, or a flow pattern that is counterflow in the central portion of the exchanger proper and crossflow near the ends.

Most of the exchangers described are of fixed type. A few are of rotary type.

Most of the exchangers described have water-impermeable heat-transfer sheets: no water is transferred from one airstream to another. A few do provide for transfer of water.

Most of the exchangers are said to have efficiencies of 60 to 80% -- and higher in a few cases. Actual efficiencies of several exchangers have been evaluated by a group (W.J. Fisk et al) of the Lawrence Berkeley Laboratory, University of California. They obtained values ranging from about 48% to 75%. See Bibl. item F-70.

Prices range from about $110 to more than $1000. Comparisons of prices may be near meaningless unless account is taken of differences in airflow rates, efficiency, and requirements as to ducts for stale air collection and fresh air distribution. Attention should be given also to differences in controls, vulnerability to frosting, durability, ease of maintenance, and many other factors discussed in previous chapters.

Chapter 19

BERNER INTERNATIONAL CORP. EXCHANGERS

The company

Range of products

Econofresher GV 120 exchanger

Senex exchangers

THE COMPANY

Berner International Corp., 12 Sixth Rd., Woburn, MA 01801. Tel.: (617) 933-2180. Fischer, John. General Manager. Eriksson, Krister. Head of engineering.

A part of the company that makes air-curtain equipment is located in New Castle, PA (PO Box 5205), 16105.

Far East sales office: Berner International Corp., Nihon Jitensha Kaikan, 9-15, 1-Chome Akasaka, Minato-Ku, Tokyo, Japan 107. Tel.: 585-6421.

European sales office: Berner International BMBH, Ausschlager Weg 71, 2000 Hamburg 26, West Germany. Tel.: 257436.

RANGE OF PRODUCTS

Econofresher GV 120, a rotary-type air-to-air heat-exchanger for houses. Made by Sharp Corp. in Japan, this exchanger was introduced into USA by Berner International Corp. in the fall of 1981.

Senex rotary-type air-to-air heat-exchangers for use in office buildings, hospitals, factories, industrial plants, etc. A wide variety of models has been marketed for many years.

ECONOFRESHER GV 120

This is a small, wall-mounted, 2-speed, counterflow, laminar flow, rotary, air-to-air enthalpy-exchanger 21½ in. x 12 in. x 8¼ in. With airflow rates of 60 cfm in each airstream, overall power consumption is 40 w and the efficiency of enthalpy recovery is about 75%. Operated at 30 cfm the exchanger uses 25 w and has an efficiency of 82%. Price: about $300 from company headquarters, or about $350 from dealer.

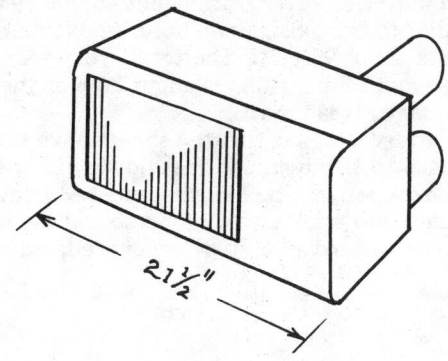

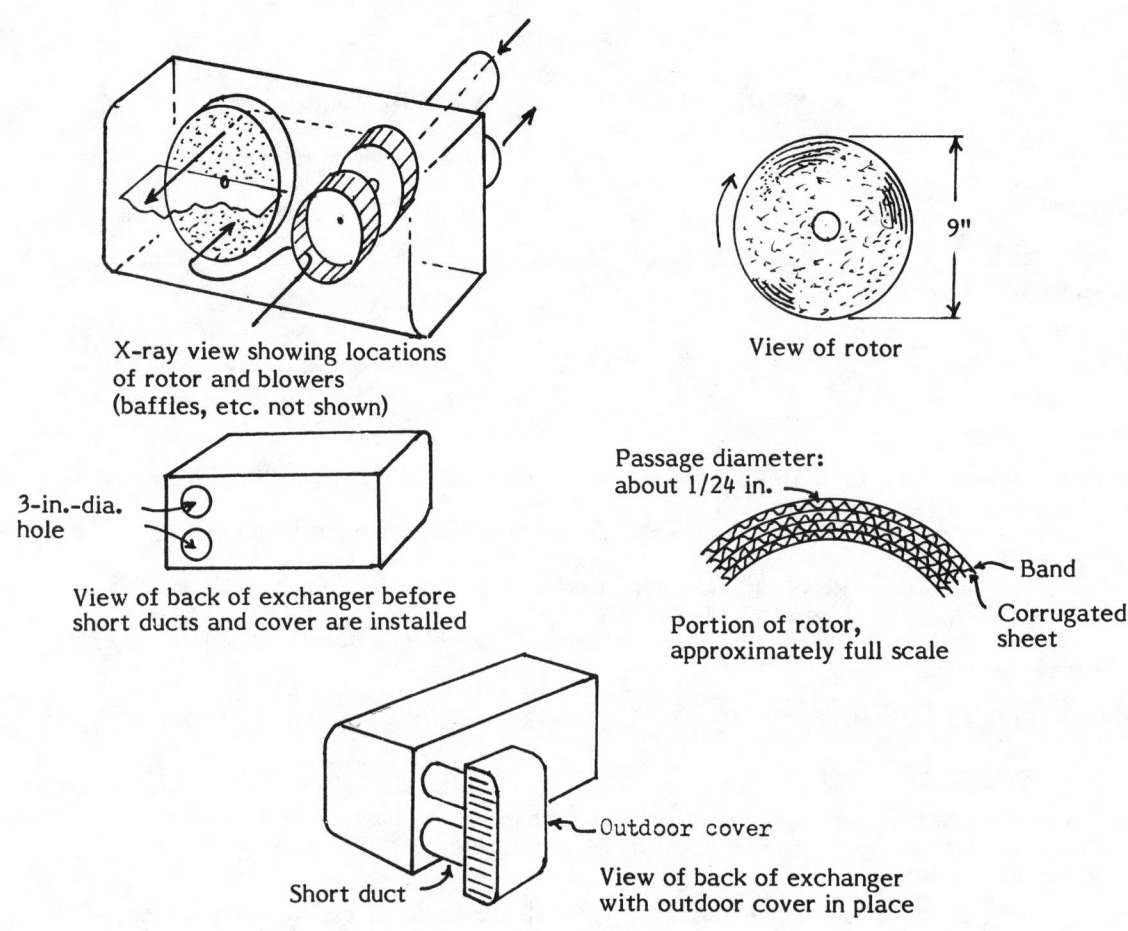

X-ray view showing locations
of rotor and blowers
(baffles, etc. not shown)

View of rotor

9"

3-in.-dia.
hole

View of back of exchanger before
short ducts and cover are installed

Passage diameter:
about 1/24 in.

Band

Corrugated
sheet

Portion of rotor,
approximately full scale

Outdoor cover

Short duct

View of back of exchanger
with outdoor cover in place

Dimensions And Weight

The exchanger is 21½ in. x 12 in. x 8¼ in. Its weight is 21 lb.

Rotor

The enthalpy-exchanger rotor is 9 in. in diameter and 2.6 in. thick. It consists almost entirely of 0.005-in.-thick plastic sheets: smooth plastic-sheet bands and intervening corrugated sheets. The plastic is DuPont's Nomex, a Teflon-base material. There are about 40,000 slender (1/24-in. dia.) parallel passages through the rotor, parallel to the axis. Each has a surface area of about 0.3 in.^{2}; thus the total heat-transfer surface area is about 90 ft.^{2}. The total cross-sectional area of the passages is about (40,000)(0.0013 in.^{2}) or about 50 in.^{2}; this is about 80% of the total face-area of the rotor (64 in.^{2}). The airflow in the passages is laminar.

All faces of the plastic bands and intervening corrugated sheets have been impregnated with a desiccant: lithium chloride (LiCl). This is so firmly bound that no significant deliquescence or migration occurs even during prolonged operation in a humid atmosphere. (However, if the rotor is removed and washed in water, the desiccant will wash away. A person whose house tends to be too humid, and who does not desire the outgoing water vapor to be recovered, may take advantage of this fact.)

The rotor rotates at 9 rpm. Its speed does not change when the blower speeds are changed.

The rotor assembly includes no "purge sector" such as is used in exchangers of much greater size. In exchangers used in houses no harm is done if a very small fraction (of the order of 0.5%) of the outgoing air finds its way into stream of incoming air.

Blowers

There are two blowers: one for each airstream. The blowers are of centrifugal (squirrel-cage) type, are mounted on a common shaft, and are driven by a single electric motor. This provides two choices of speed: high and low, controlled by a pull-string at the lower right corner of the assembly. High and low speeds provide 60 and 30 cfm flowrates respectively (in each airstream) and required 40 and 25 w of electrical power.

Filters

There are two filters, easily accessible and easily cleaned.

Housing

This is of plastic and has an imitation wood (mahogany) finish.

The small outdoor cover, concealing the two short ducts that pierce the wall of the house, excludes rain, snow, leaves, etc.

Performance

Operated at 60 cfm, the exchanger has an efficiency of enthalpy transfer of about 75%. Under certain typical cold-winter conditions it recovers about 70% more enthalpy than is recovered by an exchanger that recovers sensible heat only and has an efficiency of 75% -- all as explained in Chapter 11. Operated at 30 cfm, the efficiency of enthalpy transfer is 82%.

Condensation of water does not occur except under special and extreme conditions of outdoor temperature and humidity. See Chapter 11.

Frosting does not occur except when the outdoor temperature is very low: $0^{o}F$ if room temperature and humidity are $70^{o}F$ and 50% RH and outdoor RH is below 30%; or $-10^{o}F$ if room temperature and humidity are $70^{o}F$ and 35% RH and outdoor RH is below 30%.

Defrosting Procedure

Turn exchanger off for an hour or two.

Summer Use

The exchanger is highly effective in summer also. While bringing in fresh air it does not bring in much heat or water from the (very hot, very humid) outdoor air.

Price As Of January 1982

About $350 from dealer and about $300 direct from company headquarters. Price includes the outdoor cover.

Miscellaneous: An instruction and installation manual is included. Also a general one-year guarantee. The user should clean the filters when and if they become partially clogged, and he should inspect (and, if necessary, clean) the rotor faces several times a year.

If a user finds his house to be much too humid in winter, and therefore wishes not to recover moisture, he should remove the rotor and wash it in water -- so that it will thenceforth recover sensible heat only.

Note: In March 1982 the company introduced a ceiling-mounted exchanger, called Economini 500, that is also a rotary device made by Sharp Corp. It has greater power, 3 speeds, flowrates up to 250 cfm. Dimensions: 53 in. x 27 in. x 14 in. high. Price: $1200.

SENEX EXCHANGERS

For many years the company's main product has been air-to-air heat-exchangers that are very large and are suitable for use in office buildings, hospitals, factories, and industrial plants. Some of the exchangers transfer sensible heat only and others, which contain desiccant, transfer sensible heat and latent heat.

Each of these exchangers employs a rotor that is 3 to 14½ ft. in diameter and, typically, 8 in. thick. Typical material: aluminum -- fluted corrugated aluminum honeycomb with strengthening bands and steel spokes. Typical rotational speed: 10 to 20 rpm. Rotation is produced by a fractional-HP, belt-driven motor.

The rotor contains of the order of 100,000 to several million slender passages, parallel to the rotor axis, for airflow. The effective diameters of the passages are so small (of the order of 1/16 in.) that (a) the flow is laminar and (b) the total heat-transfer surface area is enormous -- 1000 to 10,000 ft^2, typically. Of the area of either face of the rotor, about 90% is open.

Two centrifugal blowers are used. A choice of speeds (flowrates) is provided.

One airstream (the incoming fresh air, say) flows through whichever half of the rotor is uppermost. The other airstream flows through the half that is below. The associated ducts lie one above the other. Seals are provided along the perimeter of the rotor and also along the horizontal diameter, on both faces. The seals are of self-adjusting type and are of a durable polymer.

A small, sector-shaped baffle, held in fixed position, serves to prevent any transfer of air from the outgoing stream to the incoming stream. Such a baffle, called a purge sector, is essential if the outgoing air contains highly toxic chemicals and the incoming air is delivered to rooms containing people. The full angular width of the purge sector is 20 degrees, typically.

Two specific classes of such (Berner) exchangers deserve mention:

Senex L exchanger Transfers sensible heat only. Smallest model is about 38 in. x 38 in. x 13 in. and weighs 200 lb. Price: $1600.

Senex E exchanger Transfers sensible heat and latent heat. The rotor material is impregnated with a stable, non-deliquescent desiccant (silica gel). Smallest model has dimensions and weight as above. Price: $1900.

The efficiency of enthalpy transfer of such devices is 70% to 90%, typically. Such exchangers are very cost effective when volume airflow rates are high.

Chapter 20

MITSUBISHI EXCHANGERS (LOSSNAY)

The company

Range of products

VL-1500 exchanger

VL-1500-Z exchanger

VL-1500-V exchanger

VL-1500-PH exchanger

VL-500-B2 exchanger

THE COMPANY

Mitsubishi Electric Corp. Lossnay Engineering Section, Nakatsugawa Works, Nakatsugawa, Gifu Pref., Japan. (Inventor of Lossnay exchangers: Mr. Masataka Yoshino.)

Marketing overseas: Group C. 2-3 Marunouchi 2-chome, Chiyoda-Ku, Tokyo.

American sales company: Mitsubishi Electric Sales America, Inc., (sometimes called Melco), 3030 E. Victoria St., Compton, CA 90221. Tel.: (213) 537-7132. Also: (800) 421-1132. Key person: Thomas, Mike.

Regional representatives: see list at end of chapter.

RANGE OF PRODUCTS

The company offers a wide range of Lossnay air-to-air heat-exchangers for use in houses, offices, industrial plants, etc. The exchangers provide flowrates from 23 to 200,000 cfm. The exchangers meant for use in houses are:

VL-1500	70 cfm, 29 lb., 22 x 15 x 12 in., $310. For general use. Mounted on wall. Most popular model.
VL-1500-Z	Like VL-1500 but modified for mounting in ceiling and connecting to outdoors via ducts. 62 cfm. $350.
VL-1500-V	Like VL-1500 but designed to be mounted with long axis vertical. $335.
VL-1500-PH	Like VL-1500 but includes a small electric pre-heater, for the fresh air stream, that, ordinarily, eliminates the danger of frosting. $525.
VL-500-B2	23.5 cfm, 6.2 lb., 13 x 10 x 9 in., $110. For use in bathroom or other small room.

Note concerning an obsolete model: In 1979 and 1980 the company marketed, in USA, a Model VL-1200 exchanger. Early in 1981 this model was withdrawn from sale in USA. It was slightly smaller than the VL-1500 exchanger.

116

VL-1500 EXCHANGER

Summary

The Mitsubishi Lossnay VL-1500 wall-mounted, 3-speed, crossflow, air-to-air heat-exchanger is 22 in. x 15 in. x 12 in., weighs 29 lb., and is said to provide 71, 50, or 28 cfm in each airstream, with an enthalpy exchange efficiency of 70, 73, or 76% respectively.

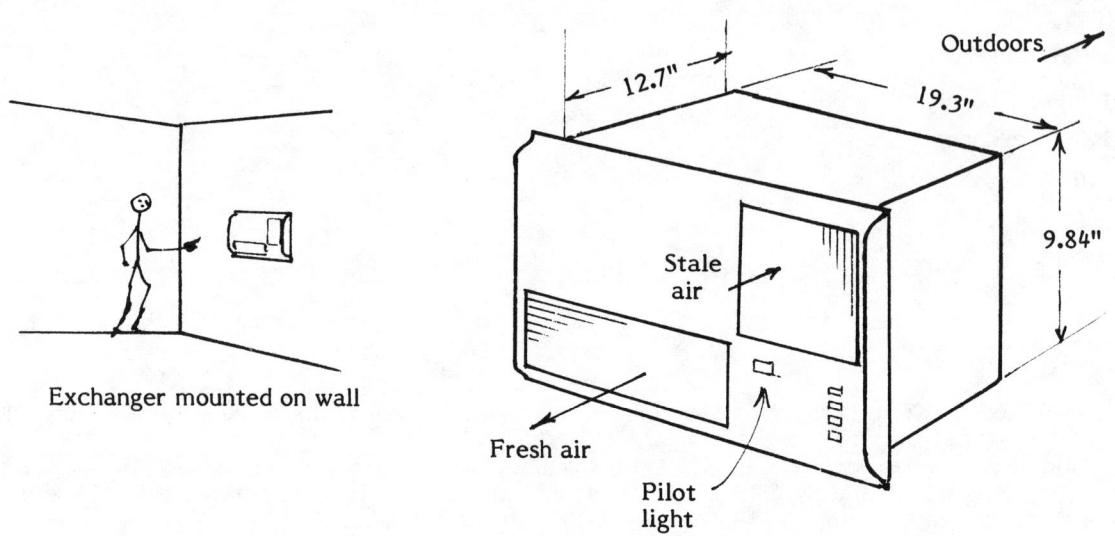

Exchanger mounted on wall

Perspective view, VL-1500 exchanger

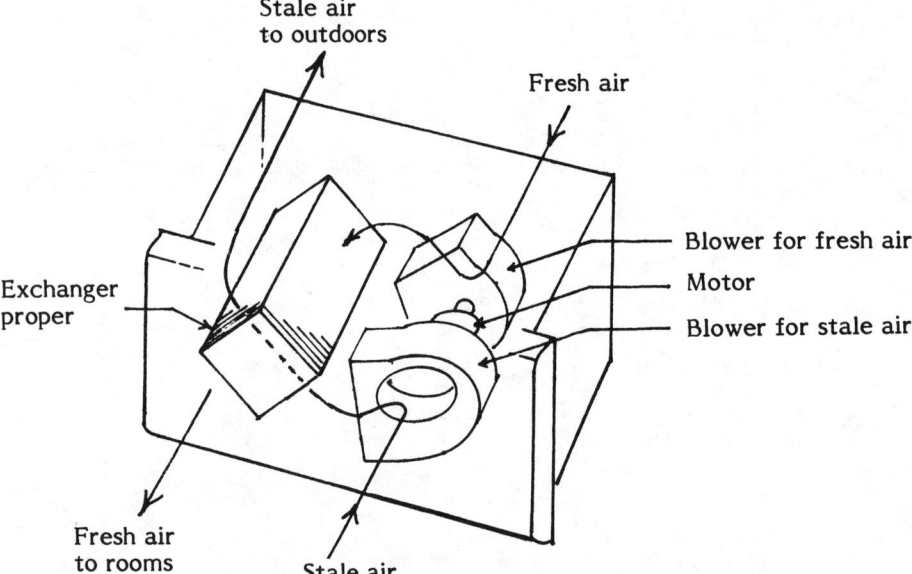

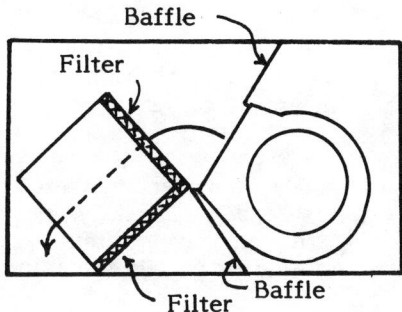

Vertical cross section through
blower serving fresh-air stream

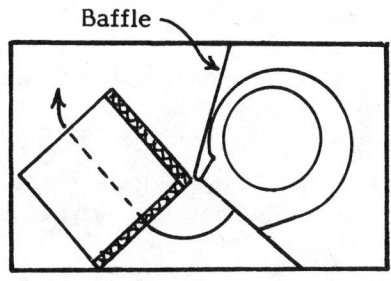

Vertical cross section through
blower serving stale-air stream

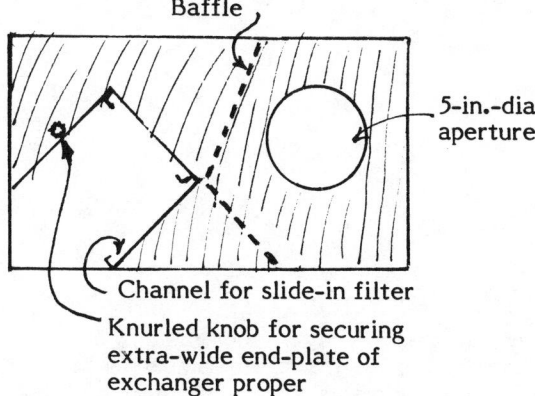

Front view of exchanger with front
panel, exchanger proper, and filters
removed.

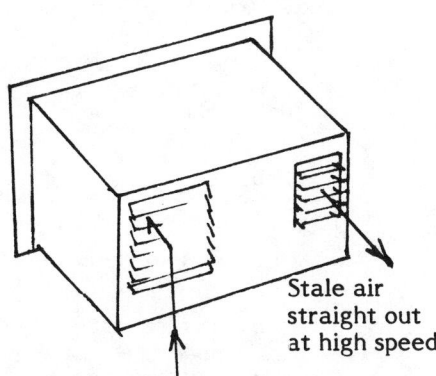

Fresh air enters in upward
direction, through steeply
slanted louvers.

Rear view of exchanger

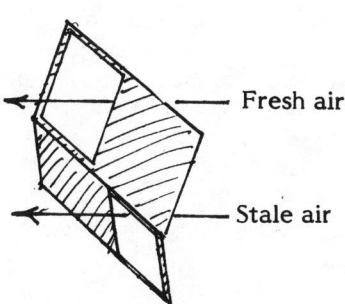

Perspective view of baffle
between blowers and
exchanger proper

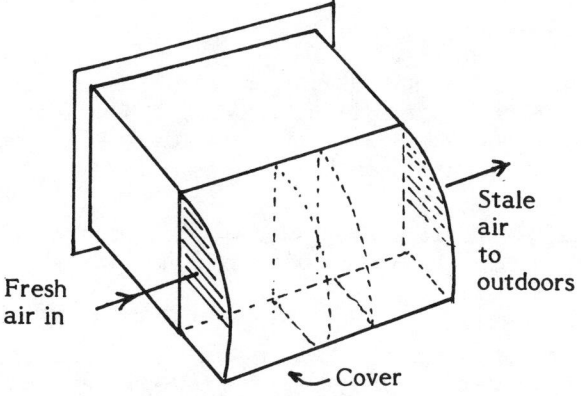

Rear view of exchanger with
cover attached

Exchanger Proper

The exchanger proper, of crossflow type, is 13 in. long and 5.1 in. x 5.1 in. cross section. It is mounted "on edge" with its long axis horizontal.

It consists of a bonded assembly of flat sheets and corrugated sheets, all consisting of specially treated paper that is strong and somewhat permeable to water. There are 90 flat sheets, spaced 1/7 in. apart on center. Interleaved between them, and bonded to them, are 89 corrugated sheets, with alternate sheets having 90-deg.-different direction. The corrugations are 1/7 in. deep and there are six of them per inch. Thus between each pair of flat sheets there are 31 slender channels, or air passages, each with an effective diameter of about 1/10 in. Incoming fresh air travels within the passages associated with corrugated sheets 1, 3, 5, etc., while outgoing stale air travels within the passages associated with corrugated sheets 2, 4, 6, etc.

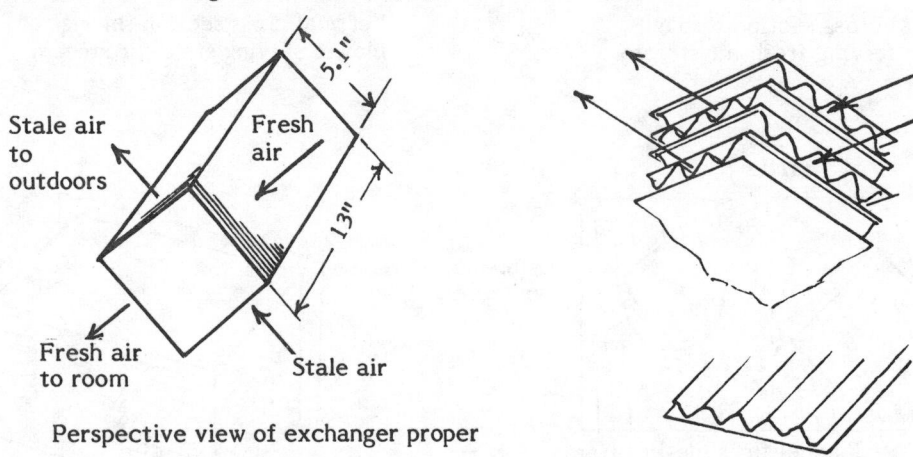

Perspective view of exchanger proper

The total surface area (of flat sheets and corrugated sheets) contacted by the outgoing air is about 40 ft^2. The corrugated sheets contribute significantly to heat-transfer inasmuch as they act as fins; the thermal conductivity of the special paper is not comparable to that of metal, but is an order of magnitude greater than that of the airfilms at the boundaries of laminar-flow air.

Because the sheets are of material that is somewhat permeable to water, some moisture from the outgoing air may enter the sheets, migrate through them (the flat sheets, that is), and evaporate into the stream of incoming (less humid) air. I have been told that about 40%, typically, of the moisture (and latent heat) in the outgoing air is recovered.

Blowers

There are two blowers: one for each airstream. Each is of squirrel-cage (centrifugal) type, 9 in. in diameter. They are mounted on a common axis and are driven by a single motor mounted coaxially with them. Operated at high, medium, or low speed (drawing 56, 40, or 22 watts), each blower provides an airflow of 71, 50, or 28 cfm respectively.

Filters

In each airstream, there is a 12 in. x 5 in., screen-like, fiberglass filter mounted upstream from the exchanger proper. Mesh: about 1/16 in.

Control Switch

This switch, situated in lower right corner of panel, has four positions: high, medium, low, and off. It is operated by pulling down on a 1/16-in.-diameter cord hanging down from the right end of the housing.

Miscellaneous Components

A standard (60 cycle, 12-v.) cord and plug are provided. There is one pilot light -- to show whether power is on. There is no fuse. An optional ($22) accessory is a protective cover, of painted weather-proof steel, that may be installed on the outdoor side of the exchanger to protect it from rain and snow and to further separate the two airstreams there. The inlet and outlet grills of the cover are in the ends thereof.

Choice Of Color: There are two choices of color of front of housing: wood grain (like mahogany) and cream. The respective designations are VL-1500-M and VL-1500-C.

Disposal of Condensate: Little condensation is expected, normally, and no special provision for disposal of condensate has been made. In some situations some dripping of water to the outdoors occurs.

Defrosting: Frost may form inside the exchanger when the outdoor temperature is below 18°F (?). To defrost the exchanger, turn it off for an hour (?).

Performance: Running at normal (high) speed, the exchanger provides 71 cfm airflow in each air-stream and the efficiency of sensible-heat-recovery is 70%. The values for medium-speed operation are 50 cfm, 73%, and the values for low-speed operation are 28 cfm, 76%.

Operation: Just plug it in, turn switch to desired speed (high, medium, or low) and let it run. If the electric power supply fails, the exchanger will stop, without damage. When the supply is reinstated, the exchanger will start up again, without harm.

Summer Use: The exchanger can be used to advantage in summer -- on days when the indoor temperature is 70 to 80F and the outdoor temperature is much higher than this. Now, thanks to the fact that the exchanger surfaces are permeable, the exchanger tends to expel excessive moisture normally present in the incoming (hot and humid) outdoor air.

Retail Price As Of January 1982: $310 for basic exchanger, plus about $22 for outdoor cover, if this is desired, plus about $15 shipping charge.

Delivery Period: Three weeks. U.S. Customs duty: None.

Instruction And Installation Manual: A 4-p. manual is provided.

Warranty: Installation dimensions are warranted.

Maintenance: About twice a year the exchanger proper and the filters should be vacuum cleaned. To remove front panel of exchanger, pull bottom forward (so that knife-type spring-clip releases) then move entire panel upward and forward. To remove exchanger proper, unscrew knurled knob and pull exchanger proper forward, out of the housing. Filters slide out forward. All of these removal operations can be performed in a half-minute without use of tools.

VL-1500-Z EXCHANGER

This exchanger is much like the VL-1500 exchanger described above, but is modified for mounting in the ceiling. Typically, 5-in.-diameter ducts (procured and installed by the purchaser) are used to carry the two airstreams through the attic. Dimensions: 23 in. x 16½ in. x 12 in. Weight: 29 lb. Airflow rates: 62 and 40 cfm, with efficiencies of 60 and 64%. Price: $350.

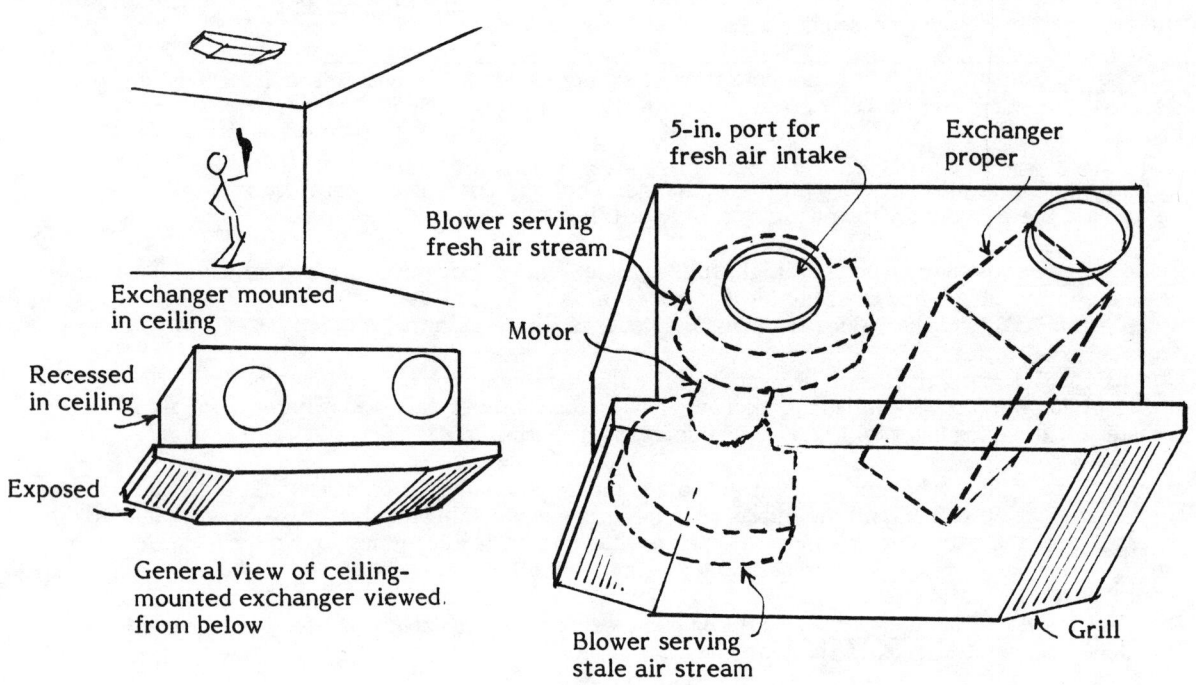

Exchanger mounted in ceiling

Recessed in ceiling

Exposed

General view of ceiling-mounted exchanger viewed from below

Perspective X-ray view, highly simplified

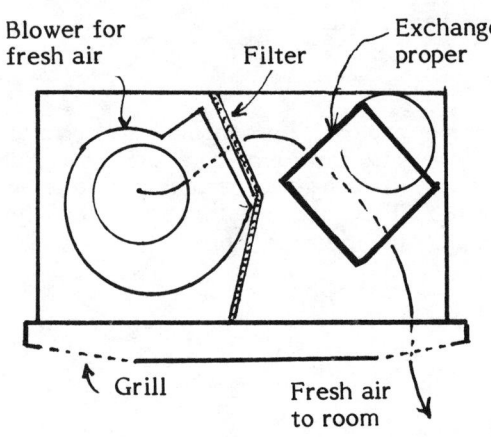

Vertical cross section through blower serving fresh air stream

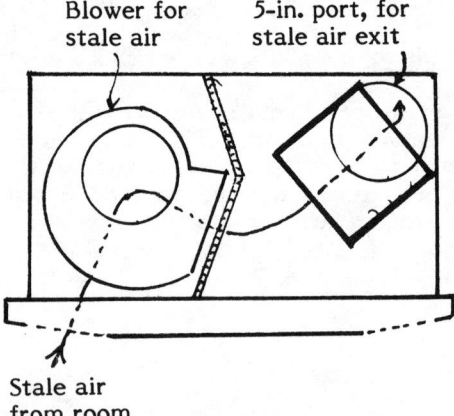

Vertical cross section through blower serving stale air stream

VL-1500-V EXCHANGER

This is like VL-1500 but is designed to be mounted with the long axis vertical. $335.

VL-1500-PH EXCHANGER

This is like the VL-1500 but includes a small electric pre-heater, for the fresh air stream. The pre-heater prevents frost formation under nearly all cold-winter conditions. $525.

VL-500-B2 EXCHANGER

This exchanger, intended to serve a bathroom or other small room, is a crossflow, one-speed device 13 in. x 10 in. x 9 in., weighing 6.2 lb. With 23.5 cfm flowrate, it provides sensible heat recovery of 65%. No latent heat is recovered.

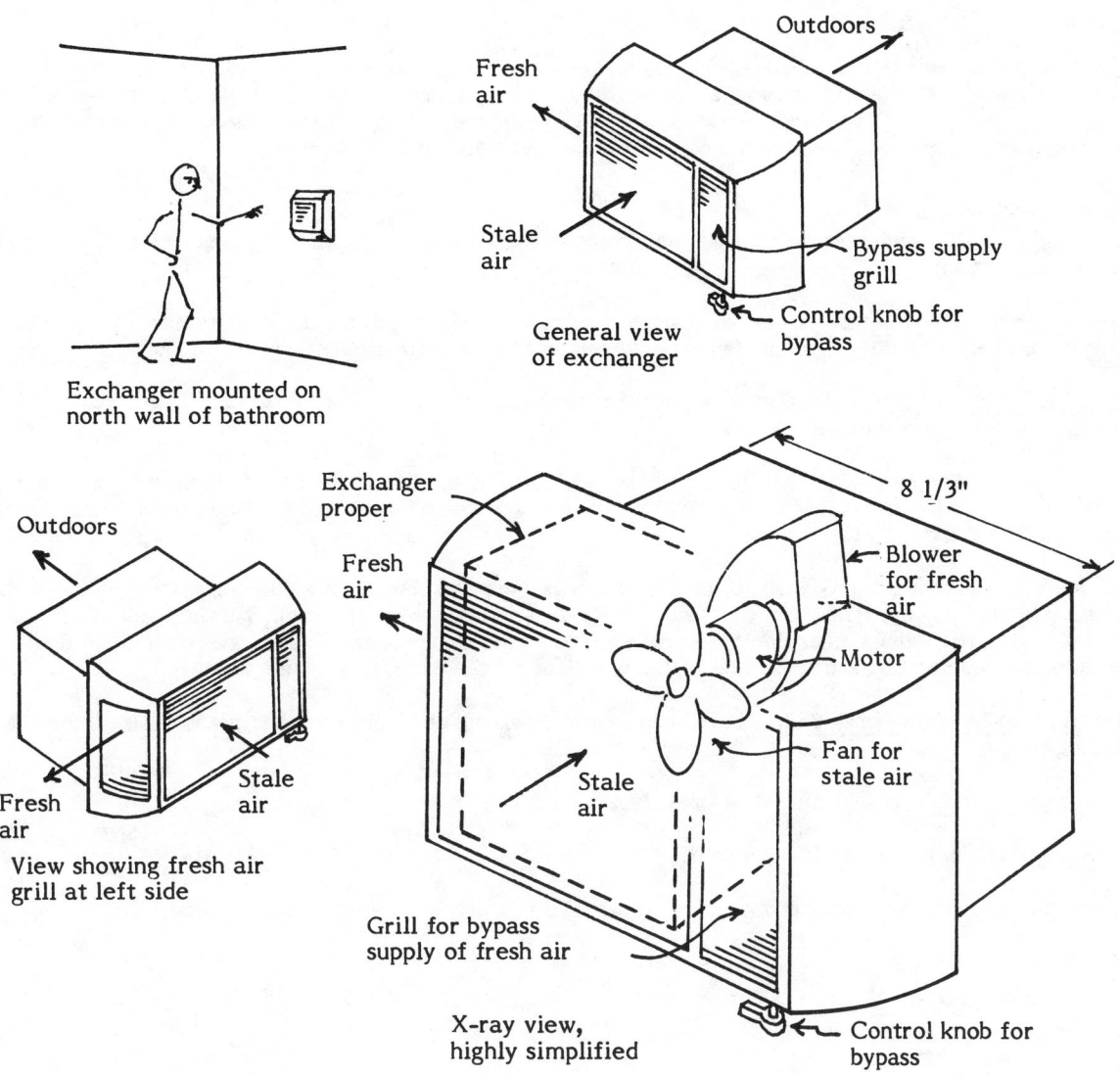

Exchanger mounted on
north wall of bathroom

General view
of exchanger

View showing fresh air
grill at left side

X-ray view,
highly simplified

Exchanger Proper

Like the VL-1500 exchanger proper except smaller (8¼ in. x 4 in. deep by 6½ in. high) and oriented with all channels horizontal. The material used is <u>not</u> permeable to water -- bathroom humidity is often high enough so that recovery of water would not be welcome.

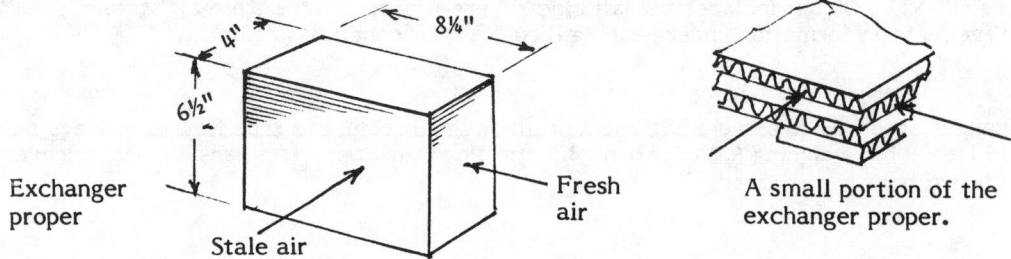

Exchanger proper   Stale air   Fresh air   A small portion of the exchanger proper.

Blowers

There is one fan (close to exchanger proper) and one centrifugal-type blower (mounted farther back). These two, and also the 30-w motor, are mounted on a common shaft that is horizontal and perpendicular to the wall. The fan drives the stale air and the blower drives the fresh air. Suitable baffles are provided to keep the two airstreams separate. There are no filters.

Miscellaneous Components

Cord and plug of standard type. Cord-type simple pendant on-off switch. No fuse or pilot light. Color of housing: cream.

<u>Condensate and frosting</u>: Outgoing air carries any condensate to outdoors. Inasmuch as the exchanger is operated for brief intervals only (when installed in a bathroom), frosting seldom occurs.

<u>An accessory: a cover</u>: A protective cover of plastic may be purchased for about $20 and may be left in place summer and winter.

<u>Performance</u>: The exchanger is said to provide 23.5 cfm of airflow in each airstream, with a sensible-heat-recovery efficiency of 65%. The exchanger does <u>not</u> recover moisture -- because the heat transfer material is impermeable.

<u>Summer</u>: The occupant may wish to turn the control knob and thus open a damper associated with the grill for bypass supply. This permits the incoming fresh air to enter the room via the (special) small grill -- instead of being constrained to enter via the exchanger proper. Thus more fresh air enters. Because outdoor air is warm in summer, the bypassing of the exchanger is not wasteful.

<u>Retail price as of January 1982</u>: $110 plus about $20 for the outdoor cover, if this is desired, plus $6 for shipping.

<u>Instruction manual</u>: A 4-p. manual is provided.

<u>Warranty</u>: Covers all parts for one year.

<u>Maintenance</u>: The exchanger proper must be washed in water about every six months.

Regional Representative for
Mitsubishi Lossnay Exchangers

Region	Representative
Alaska	The Energy Store, 3934 Spenard Rd., Anchorage, AK 99503. (907) 248-1948
Arkansas (south)	Same as Mississippi
California (south)	Haldeman Inc., 2845 Supply Ave., Los Angeles, CA 90040. (213) 726-7011
California (north bay area)	Western Mech Co., 3402M Mt. Diabolo Blvd., Lafayette, CA 94549. (415) 284-1310
California (north)	Westates Equipment, 8670 23 Ave., Sacramento, CA 95826. (916) 381-2552
Colorado	Pearsall Co., 1030 W. Ellsworth, Denver, CO 80223. (303) 573-1918
Connecticut	Same as Massachusetts
Delaware	Same as New Jersey
Florida	Energy Management Control Inc., 1279 Kingsley Ave., Suite 109, Orange Park, FL 32073. (904) 269-5005
Georgia	Bache Industrial Pr., 3016 Adriatic Ct., Norcross, GA 30366. (404) 449-0680
Hawaii	AMFAC Distribution Co., Ltd., 694 Auahi St., Honolulu, Hawaii 96827. (808) 533-0367
Idaho (north)	Same as Washington
Idaho (other)	Same as Utah
Illinois (Champaign)	Solar Design Associates, Inc., 205 West John, Champaign, IL 61820. (217) 359-5748
Illinois (Chicago)	ESP Associates, 1107 S. Mannheim, Westchester, IL 60153. (312) 343-2882
Illinois (other)	Natural Energy Systems, Inc., 3407 N. Ridge, Arlington Heights, IL 60004. (312) 359-6760
Illinois and Missouri	Caruso Sales, 558 N. Cherry St., Galesburg, IL 61401. (309) 343-8707
Indiana (central)	McIver Equipment, 1111 E. 54 St., Indianapolis, IN 46220. (317) 255-3606
Indiana (northern)	Enviro Therm, 8128 Jackson St., Munster, IN 46321. (219) 924-0496
Iowa	Conservation Engineering Ltd., 301 E. Walnut, Des Moines, IA 50309. (515) 243-6472
Louisiana	Energy Concepts, 5360 Keele St., Jackson, MS 39206. (601) 981-6649
Maine	Same as Massachusetts
Maryland	Same as Pennsylvania (eastern)

124

Massachusetts	Solar Alternative Inc., 71 Main St., Brattleboro, VT 05301.
Michigan	Ramsey Air Conditioning, 2000 Turner St., Lansing, MI 48906. (517) 482-3710
Minnesota	EER Products Inc., 4501 Bruce Ave., Minneapolis, MN 55424. (612) 926-1234
Missouri	Brusco-Rich Inc., 10621 Liberty Ave., St. Louis, MO 63132. (314) 428-2900
Montana (western)	General Sales Co., Inc., West 408 Indiana, Spokane, WA 99205. (509) 328-8600
Montana (other)	ABCO Supply Inc., 4005 1st Ave. South, Billings, MT 59107. (406) 248-7808
Ohio	Columbus Temperature Corp., 1053 E 5th Ave., Columbus, OH 43201 (614) 294-6216
Oregon	Interstate Specialty Products Inc., 5708-B SE 135 Ave., Portland, OR 97236. (503) 761-2199
Nevada (western)	Same as California (northern)
New Hampshire	Same as Massachusetts
New Jersey	Renner Energy Industries, PO Box 217, Oaklyn, NJ 08107. (609) 848-7488
New York (southeast)	BTU Systems, 451 Fulton Ave., Hempstead, NY 11550. (516) 481-3344
North Carolina	AE Products, 655M Pressley Rd., Charlotte, NC 28203. (704) 527-4282
North Dakota	Same as Minnesota
Pennsylvania (eastern)	Energy Technics Co., PO Box 3424, York, PA 17402. (717) 755-5642
Pennsylvania (western)	Energy Technics, 804 Mt. Hood Dr., Pittsburgh, PA 15239. (412) 824-7606
Rhode Island	Same as Massachusetts
South Carolina	Same as North Carolina
South Dakota	Same as Minnesota
Utah	AA Maycock Co., 336 W. 700 South St., Salt Lake City, UT 84110. (801) 364-1926
Vermont	Same as Massachusetts
Virginia	Energy Systems Analysis, Box 157 F, White Post, VA 22663. (703) 837-2988
Washington	Energy Devices, PO Box 717, Everett, WA 98206. (206) 252-2166
Wisconsin	LaCrosse Plumbing Supply Co., 106 Cameron Ave., LaCrosse, WI 54601. (608) 784-3839
Wyoming	Same as Colorado
Canada, Ontario	Advanced Idea Mechanics Ltd., 26 Ski Valley Crescent, London, Ont. N6K 3H3, Canada. Tel.: (519) 471-5573

Chapter 21

Q-DOT EXCHANGERS

The company

Range of products

The TRU 120M-6A exchanger

The TRU 120M-6B exchanger

The TRU 120-12-36-5-14-AC5 exchanger

Other Q-Dot exchangers

THE COMPANY

Q-Dot Corp., 726 Regal Row, Dallas, TX 75247. Tel.: (214) 630-1224. Bucher, Axel. Manager of International Marketing. Bullard, Mike. Manager of National Marketing.
Regional representatives: list is available from company.

RANGE OF PRODUCTS

The company's focus is on applications of heat-recovery equipment to commercial and industrial companies. They supply such equipment to home-owners also.

The heart of their equipment is a set of heat-pipes. Larger exchangers employ 1-in.-dia. heat-pipes, and smaller exchangers employ 5/8-in. dia. heat-pipes.

A typical exchanger, as supplied by Q-Dot Corp., consists of a set of parallel heat-pipes, a set of slightly corrugated sheets (called fins) that are perpendicular to, and pierced by, the heat-pipes, a partition between the two halves of the assembly (to keep the two airstreams separate), and a frame and/or housing. (Note: TRU stands for Thermal Recovery Unit.)

Not included are the blowers, or sensors and controllers to turn the blowers on and off or to regulate their speeds. Likewise, the necessary input and output ducts for fresh air and stale air are not supplied. All these must be provided by the purchaser.

A company spokesman said that these heat-pipe-type exchangers, although having a higher first cost than some competing exchangers of more conventional type, tend to have a lower pressure drop, employ blowers of lower power, and provide higher efficiency of sensible heat recovery than many of these competing types of exchangers.

Because the heat-pipes are horizontal, in standard installations, they perform equally well in winter and summer. In winter they help keep heat in, and in summer they help keep it out.

THE TRU 120M-6A EXCHANGER

This exchanger, designed for use in average-size homes, employs a total of 24 heat-pipes arranged in six rows of 4 heat-pipes each. The overall dimensions of the exchanger are: 22.8 in. long x 9 in. high x 9.3 in. thick. Weight: 100 lb. The net region containing heat-pipes and fins has these dimensions: 18.5 in. long x 9 in. high x 7.8 in. thick. There is a central vertical partition (parallel to the two airstreams, which are side-by-side and horizontal) of 20 gauge galvanized steel. The entrance aperture for each airstream is 9.3 in. long x 9 in. high.

126

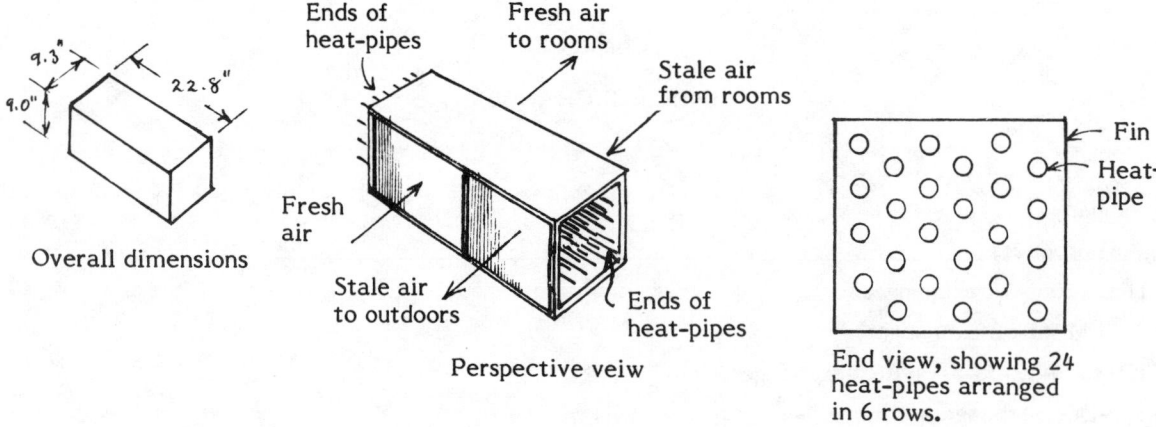

Ends of heat-pipes

Fresh air to rooms

Stale air from rooms

Fresh air

Stale air to outdoors

Ends of heat-pipes

Overall dimensions

Perspective veiw

Fin

Heat-pipe

End view, showing 24 heat-pipes arranged in 6 rows.

Each heat-pipe is of aluminum, is about 22 in. long and 5/8 in. in diameter. It contains a special fluid and a wick and is entirely sealed. The fluid, which is basically a refrigerant, continually evaporates at the hotter end of the heat-pipe and condenses at the cooler end. The condensate returns to the hotter end -- by gravity. There are no motors or pumps or other mechanical moving parts. Only the fluid moves. The evaporation process absorbs heat from the stream of hot air, and the condensation of the vapor delivers heat to the stream of cold air.

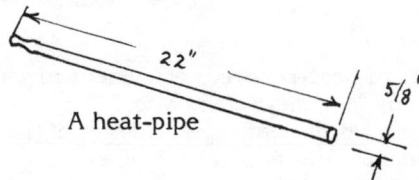

22"

5/8"

A heat-pipe

The exchanger presents, to each airstream, an entrance aperture area of 0.58 sq. ft.

Each fin is of 0.009-in. aluminum and is large enough so as to serve all of the tubes, i.e., is pierced by all of them. Each fin is corrugated; the corrugations are shallow (1/16 in. or less). There are 14 fins per inch and the airspaces between them are about 0.07 in. thick. The airflow in these spaces is laminar.

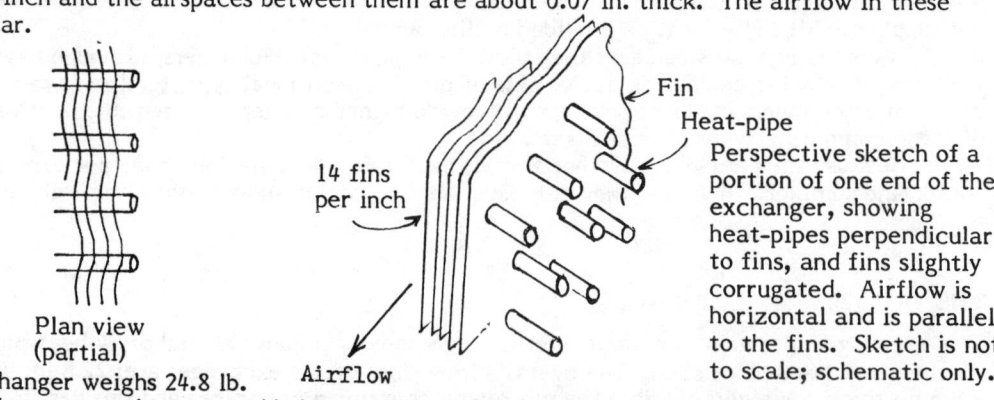

14 fins per inch

Fin

Heat-pipe

Perspective sketch of a portion of one end of the exchanger, showing heat-pipes perpendicular to fins, and fins slightly corrugated. Airflow is horizontal and is parallel to the fins. Sketch is not to scale; schematic only.

Plan view (partial)

The exchanger weighs 24.8 lb.

Airflow

The exchanger may be mounted in basement, or elsewhere indoors or outdoors. The overall housing, and the sets of ducts for input air and output air, and also the motors, blowers, and controls, are to be provided by the purchaser. The ducts may be circular or rectangular in cross section. The blowers may be upstream or downstream from the exchanger and may be close to it or remote from it. The fresh air may be supplied to one room or (via a set of ducts) to many rooms.

Under typical conditions, and with an airflow rate of 230 cfm in each airstream, corresponding to a linear speed of 400 ft/min. and a 0.58 ft^2 entrance aperture of exchanger proper, the efficiency of sensible-heat-recovery is about 64% and the pressure drop in the exchanger proper is 0.47 in. of water.

The retail price of such exchanger may be expected to be about $400, not including, of course, the components to be supplied by the purchaser.

The exchanger is covered by several patents.

It has a one-year warranty on materials and workmanship.

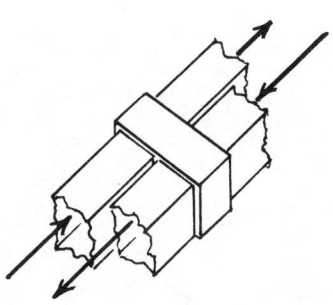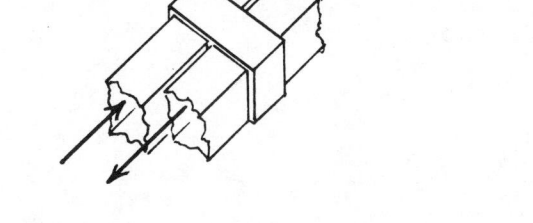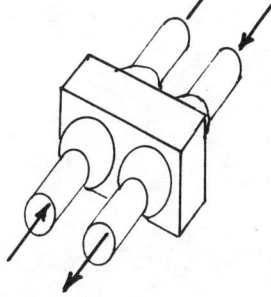

Exchanger and rectangular ducts   Exchanger and circular-cross-section ducts

THE TRU 120-M-6B EXCHANGER

This is like the above-described exchanger, but is slightly smaller. Mainly it is less high (6 in. instead of 9 in.). Its overall dimensions are 23.5 in. x 6 in. high x 9.3 in. thick. Weight: 17 lb.

THE TRU 120-12-36-5-14-AC5 EXCHANGER

This model is representative of the smaller models made for commercial applications. However, it employs 5/8-in.-diameter heat-pipes, whereas most of the exchangers for commercial use employ 1-in.-diameter heat-pipes.

It is much like the TRU 120M-6A exchanger, but has top, bottom, and end cover plates. These are of 14 gauge galvanized steel (casing) and 20 gauge galvanized steel (end cover). The partition is of 0.090-in. aluminum alloy. Overall dimensions: 36 in. long x 14 in. high x 12 in. thick. There are five rows of heat-pipes.

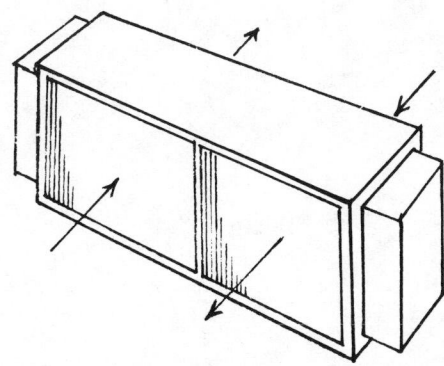

128

OTHER Q-DOT EXCHANGERS

Exchangers can be provided with a great variety of dimensions (length, height, thickness), determined by choice of number of rows of heat-pipes and number of heat-pipes in each row. Also, several exchangers can be ganged together to provide any desired shape and area of rectangular aperture for airflow. Many accessories and options are available.

A technical bulletin QHV 81-1 presents detailed information on many models. Included are graphs, nomographs, etc., to be used in predicting the performance of any exchanger in any given situation.

Chapter 22

DES CHAMPS EXCHANGERS (Z-DUCT)

The company

Range of products

Z-Duct 79M.4-RU exchanger

Z-Duct 79M.2-RU exchanger

Z-Duct 79M.6-RU exchanger

Z-Duct 74, 1000 68A6 exchanger

Regional representatives

THE COMPANY

Des Champs Laboratories Inc., PO Box 384 (also PO Box 440) East Hanover, NJ 07936. Tel.: (201) 884-1460.

Des Champs, Dr. Nicholas H. President. Wyckoff, John T., Jr. Production Manager for Residential Equipment. (Des Champs is pronounced "Day Shumps").

Regional representatives: see list at end of chapter.

RANGE OF PRODUCTS

Exchangers for use in houses: Z-Duct 79M.4-RU, now in routine production
                                       Z-Duct 79M.2-RU ⎫
                                       Z-Duct 79M.6-RU ⎭ Ready early in 1982

Exchangers for use in large buildings, factories, etc.: Assemblies of one or more modules of Z-Duct 1000 cfm. Typical module: Series 74, 1000 68A6.

Accessories: Large variety available for indirect evaporative coolers, automatic washing devices, defrosting systems, etc.

Z-DUCT 79M.4-RU EXCHANGER

Summary

The 79M.4-RU exchanger is of counterflow type and is intended to be suspended beneath a floor or installed above a ceiling. Dimensions: 5 ft. x 16 in. x 18 in. Weight: 75 lb. Flowrate: 150 cfm in each airstream.

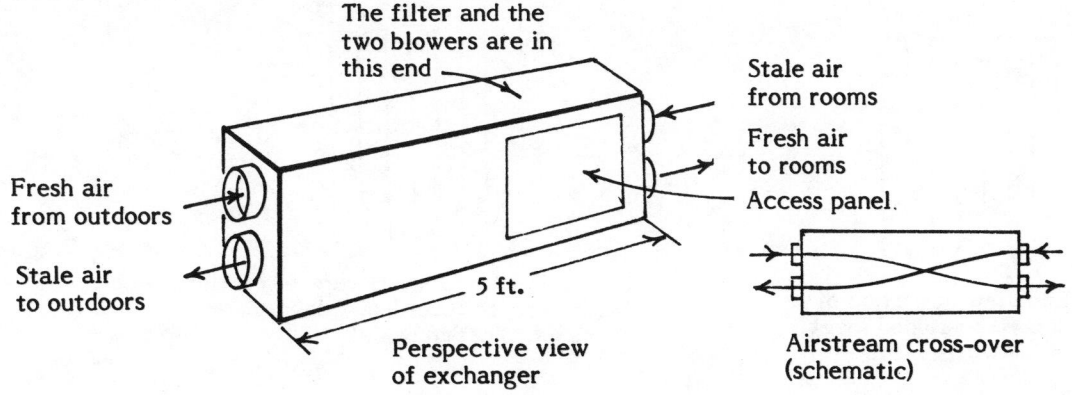

Perspective view
of exchanger

Airstream cross-over
(schematic)

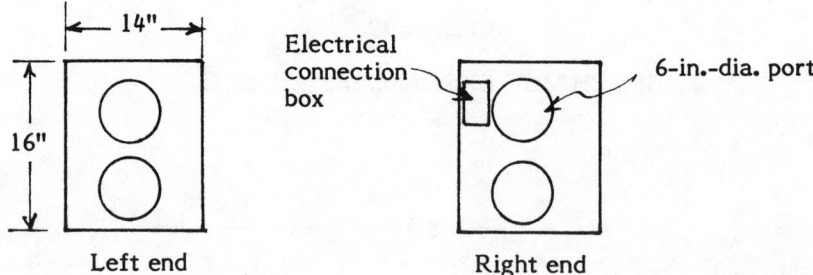

Left end

Right end

Electrical connection box

6-in.-dia. port

Exchanger Proper

The heart of the exchanger proper is a continuous sheet of 0.006-in. aluminum that has been accordion-folded and crimped, with four folds per inch. The intervening airspaces are thus about ¼ in. thick. The overall dimensions of the accordion-folded-and-crimped assembly are: 36 in. x 16 in. x 12 in. Total area of heat-exchange sheet: about 186 ft^2; the total surface area (including both sides of sheet) is about 372 ft^2.

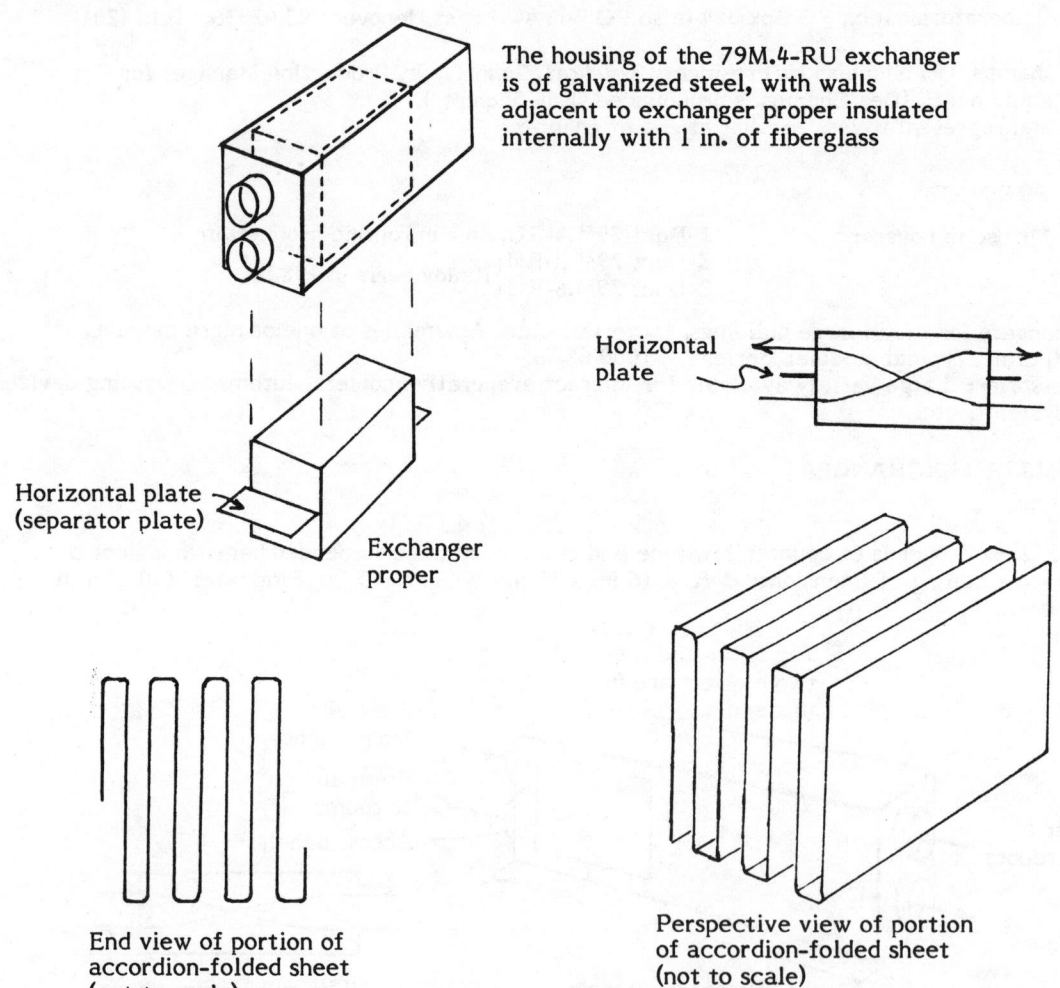

The housing of the 79M.4-RU exchanger is of galvanized steel, with walls adjacent to exchanger proper insulated internally with 1 in. of fiberglass

Horizontal plate

Horizontal plate (separator plate)

Exchanger proper

End view of portion of accordion-folded sheet (not to scale)

Perspective view of portion of accordion-folded sheet (not to scale)

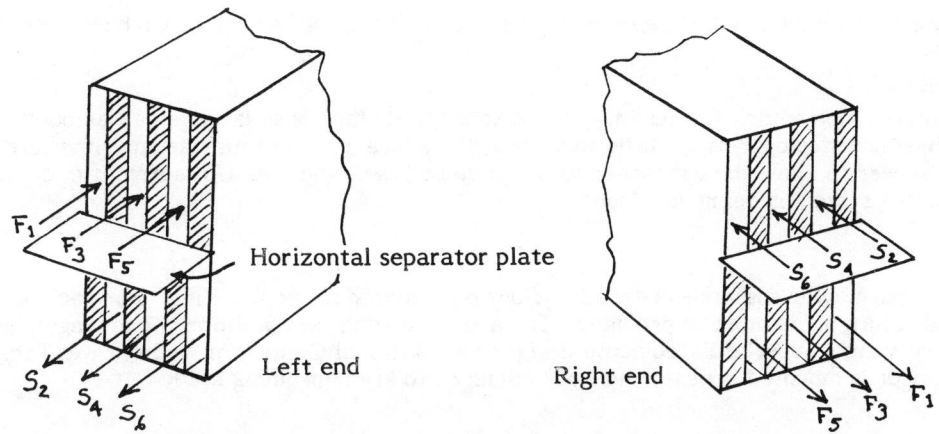

Perspective views of left and right ends of exchanger proper, showing alternating flow channels blocked off. The fresh air, traveling to the right, travels in Airspaces 1, 3, 5; it enters the upper left portion of the exchanger and leaves by the lower right portion. The stale air, traveling to the left, travels in Airspaces 2, 4, 6; it enters the upper right portion and leaves by the lower left portion. The airspaces are far more numerous, and thinner, than indicated.

In fact, there are about 50 airspaces in all. A more-nearly-to-scale drawing of the cross section is shown at the right.

The blanked-off areas are blanked off by means of a crimping operation and application of sealant.

Filter

There is a 2-in.-thick washable filter at output end (room end) of fresh air stream.

Blowers

There are two blowers, each of centrifugal type. Each uses 120 watts when operated in free air and uses 190 watts when producing a flow of 150 cfm against a pressure head (due to ducts etc.) of 0.35 in. water.

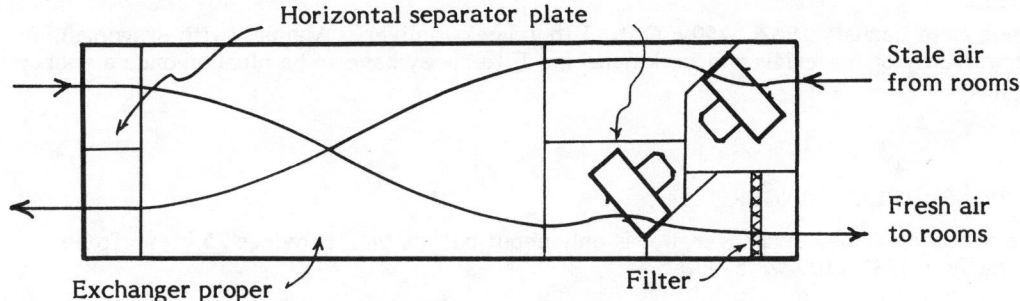

Vertical cross section showing the locations (and 45-deg. orientations) of blowers and location of the filter

Drains

Condensate that forms may escape by either of two drains, at either end of exchanger proper.

Defrosting

No special provision made. Because the heat-exchange surface is so large, small amounts of frost formation do not stop operation. If he so wishes, the house occupant may at any time turn off the fresh-air blower to allow the exchanger to warm up and melt the ice. (An automatic ice-sensing and defrosting system is being developed.)

Installation

The exchanger may be suspended beneath a floor or mounted above a ceiling. For each airstream, supply and return ducts must be provided. The accompanying sketch shows the arrangements of ducts used in a very tight superinsulated house designed by National Center for Appropriate Technology. The exchanger is mounted beneath the floor adjacent to kitchen-dining area.

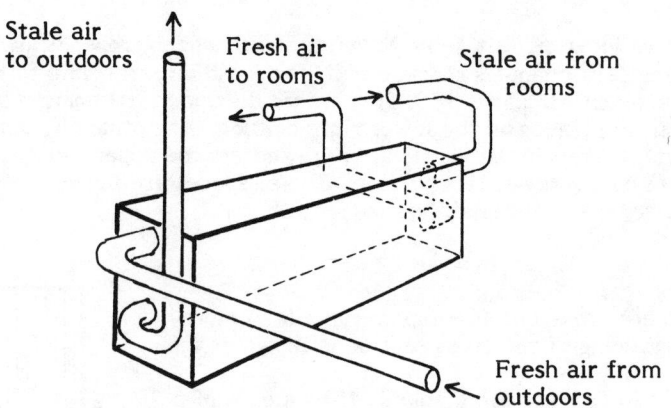

Arrangement of ducts in beneath-floor
installation in NCAT Micro-Load House

Performance

When flowrate in each airstream is 150 cfm, efficiency of sensible-heat-recovery is 85%.

Summer Use

Device can be helpful in summer also. The outgoing cool stale air cools the incoming hot fresh air. (Note: Although the company makes and sells indirect evaporative cooling equipment, no such equipment is included in this air-to-air heat-exchanger).

Miscellaneous

Retail price as of January 1982: $450 F.O.B. 3 to 4 weeks delivery. Manual (with drawings) included. One year warranty on materials and workmanship. Filter may have to be cleaned once a year with soap and water.

Z-DUCT 79M.2-RU EXCHANGER

Much like above-described exchanger, but is only about half as big. Provides 75 cfm. To be available early in 1982. 100 w. $400.

Z-DUCT 79M.6-RU EXCHANGER

Much like the 79M.4-RU exchanger, but is larger: 69 in. x 16 in. x 20 in. Flowrate: 350 cfm. Weight: 100 lb. Efficiency 85%. 380 w. $600. To be available early in 1982.

Z-DUCT 74, 1000 68A6 EXCHANGER

Summary

This is a typical type of 1000-cfm module for use in office buildings, schools, industrial plants, etc. Many modules can be combined (ganged, bolted together) to provide nearly any desired volume of air-flow. The individual module is about 36 in. x 22 in. x 18 in.; normally it stands vertical, i.e., with the 36-in. edges vertical. The exchanger proper is of Z-duct type, with aluminum heat-transfer surfaces; the aluminum sheet is folded so as to have 6 folds per inch. The housing is of corrosion-resistant, coated, steel. Each of the four ports is rectangular: 22½ in. x 9 in. During normal operation the pressure drop in the assembly (exchanger proper, and the 180-degree bends in each of the two flow-channels) is 0.89 in. of H_2O and (if the various attached ducts interpose negligible resistance) the flowrate in each stream is then 1000 cfm.; the efficiency of sensible-heat-recovery under these conditions is 68%. With 500 and 1150 cfm flows, the efficiencies are 73% and 67% respectively. (Ref. F-70 presents tests results on a slightly modified design.)

    Not included are: blowers, supply and return ducts, sensors, controls.

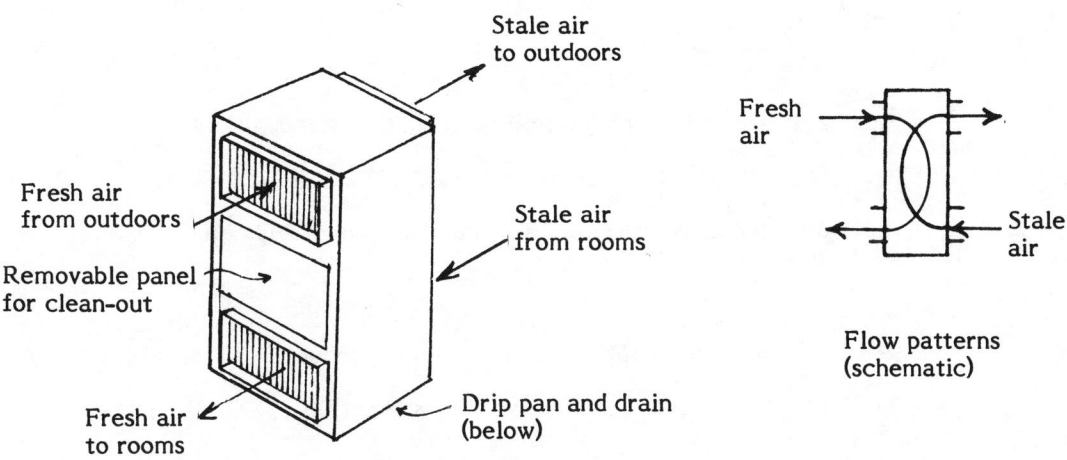

Flow patterns
(schematic)

REGIONAL REPRESENTATIVES

Region	Representative
California	Tempco Equipment, PO Box 400, Brisbane, CA 94005. Tel.: (415) 467-8680
Colorado	Conservation Mechanical Systems, 7333 S. Silverhorn Dr., Evergreen, CO 80439. Tel.: (303) 674-4634
Idaho	Sabol and Rice, Inc., 168 So. Cole Rd., Boise, ID 83705. Tel.: (208) 376-7541
Illinois	E.S.P. Assoc., Inc., 1107 S. Mannheim Rd., Westchester, IL 60153. Tel.: (312) 626-6683
Kansas	Air Systems, Inc., PO Box 636, Shawnee Mission, KS 66201. Tel.: (913) 677-1991

134

(cont'd.)

Massachusetts	New England Sunhouse, Inc., 28 Charron Ave., Nashua, NH 03063. Tel.: (603) 889-8550. Exchangers for houses.
	John Oldach Assoc., 161 Merrimac St., Woburn, MA 01801. Tel.: (617) 938-0150. Exchangers for commercial use.
Michigan	J. Pappas Assoc., Inc., 32500 Concord Dr., Madison Hgts., MI 48071. Tel.: (313) 589-1820
Minnesota	Schwab-Vollhaber, Inc 1306 W. County Rd. F, St. Paul, MN 55112. Tel.: (612) 636-3890
New Jersey	Thermo, 18 19th Ave., Paterson NJ 07513. Tel.: (201) 523-0050
New York	
Rochester	Korts Equipment Co., 306 Norton St., Rochester, NY 14621. Tel.: (716) 266-5500.
Syracuse	Korts Equipment Co., 6493 Ridings Rd., Syracuse, NY 13206. Tel.: (315) 437-0201
Tonawanda	J. Alfred Boyd & Assoc., Ltd., 90 Sunset Terrace, Tonawanda, NY 14150. Tel.: (716) 694-4423
Ohio	
Cleveland	Thermal Products Co., 1555 Hamilton Ave., Cleveland, OH 44114. Tel.: (216) 781-9796
Columbus	Pakohio, PO Box 5840, Columbus, OH 43221. Tel.: (614) 876-6311
Oregon	Howard Weller Co., 2925 NE Glisan, Portland, OR 97232. Tel.: (503) 234-5077
Pennsylvania	
Philadelphia	Robert Arnold Assoc., 1541 Sansom St., Philadelphia, PA 19102. Tel.: (215) 567-0856
Pittsburgh	J.C. Mottinger Co., 623 Center Ave., Pittsburgh, PA 15229. Tel.: (412) 931-6140
Utah	Sabol & Rice Inc , PO Box 25957, Salt Lake City, UT 84125 Tel.: (801) 973-2300
Virginia	Shultz & James, Inc., 9 E. Cary St., Richmond, VA 23219. Tel.: (804) 644-3021
Washington	
Kent	Howard Weller, 7630 S. 259 St., Kent, WA 98031. Tel.: (206) 854-9232
Spokane	Climate-Tec. Inc., N. 1108 Washington, Spokane, WA 99201. Tel.: (509) 328-4465
Wisconsin	Anguil Energy Systems, 4530 N. Oakland Ave., Milwaukee, WI, 53211. Tel.: (414) 333-0230

****

Canada	Fabrication Z-Air Corp., 690 Place Trans-Canada, Longueuil, Quebec, Canada J4G 1P2

Chapter 23

ENERCON EXCHANGERS

For information on
change in company
name and changes in
product model numbers,
see note at bottom
of page 139.

Note added in proof

The company

Range of products

Model 158 exchanger

Other exchangers

THE COMPANY

Enercon Industries Ltd., 2073 Cornwall St., Regina, Sask., Canada S4P 2K6. Tel.: (306) 585-0022.
Related companies at same address: Enercon Consultants Ltd., Enercon Building Corp., Enercon
Home Systems Corp. Key persons: Schell, Michael B. Technical Director. Wagman, Rick. Manager
of sales group. Parkes, Angela. Sales representative.

USA Subsidiary: Enercon of America, Inc., 2020 Circle Dr., Worthington, MN 56187. Tel.:
(507) 372-2442. Roslensky, David: President.

RANGE OF PRODUCTS

Four models, all called Fresh Air Heat exchanger, are available: Model 158 (150 cfm), Model 206
(200cfm), Model 408 (400 cfm), and Model 812 (800 cfm). The Model 158, sized for use in typical-
size house, is described below.

MODEL 158 EXCHANGER

Summary

The overall dimensions of this exchanger, including housing, fans, and cylindrical exchanger proper, are:
68 in. long, 20 in. diameter. Exchanger proper is 36 in. long, 20 in. in diameter. Weight: 70 lb. Airflow
is counterflow and laminar. There are about 22 concentric cylindrical plastic shells. When the fans are
operated at full power they provide 150 cfm in each of the two airstreams and the efficiency of sensible
heat recovery is 65%. When the fans are operated at low power they provide 75 cfm with 79% recovery
efficiency. January 1982 price to buyers in USA: $1495 US F.O.B. Worthington, Minn.

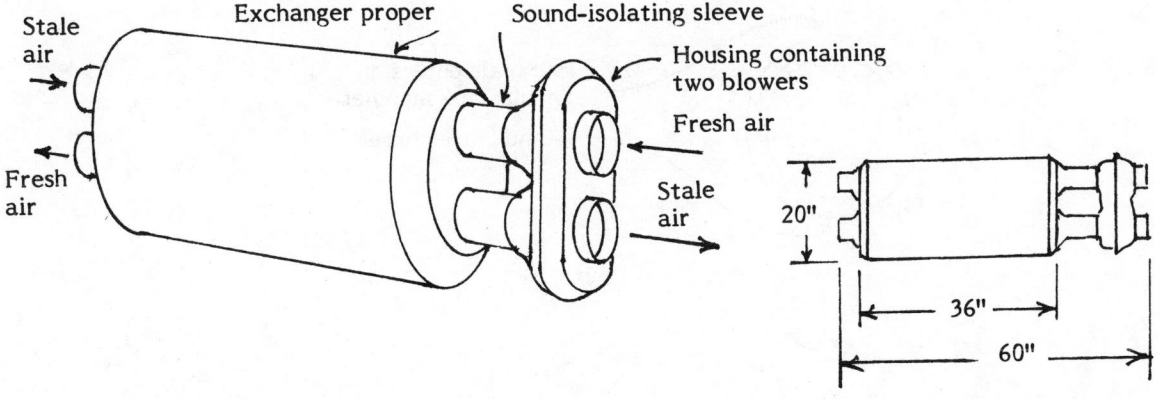

Exchanger Proper

The heart of the exchanger proper is a set of about 22 horizontal, 34-in.-long, concentric, cylindrical shells of the water-impermeable plastic polystyrene 0.020 in. thick. Diameters range from about 6 in. to about 18 in. Thus each airspace (between successive shells) is about ½ in. thick. The shells, together with the cylindrical housing, define about 23 annular-cross-section airspaces. One airstream (say, fresh air, flowing west) travels in Airspaces 1, 3, 5, etc., and outgoing stale air travels in the opposite direction in Airspaces 2, 4, 6, etc. The incoming fresh air enters the above-mentioned Airspaces 1, 3, 5, etc., via the upper portions thereof and leaves them via the lower portions; the other terminal portions are blocked off by semi-annular diaphragms or equivalent. The stale air enters its airspaces via the upper portions thereof and leaves via the lower portions. At each end of the exchanger proper there is a horizontal transverse baffle, or partition, that keeps the two airstreams separate there.

The following highly simplified diagrams explain the operation described above. It shows how the countercurrent airstreams can exchange heat (through the thin walls of the shells) while always being kept separate; no cross-contamination can occur.

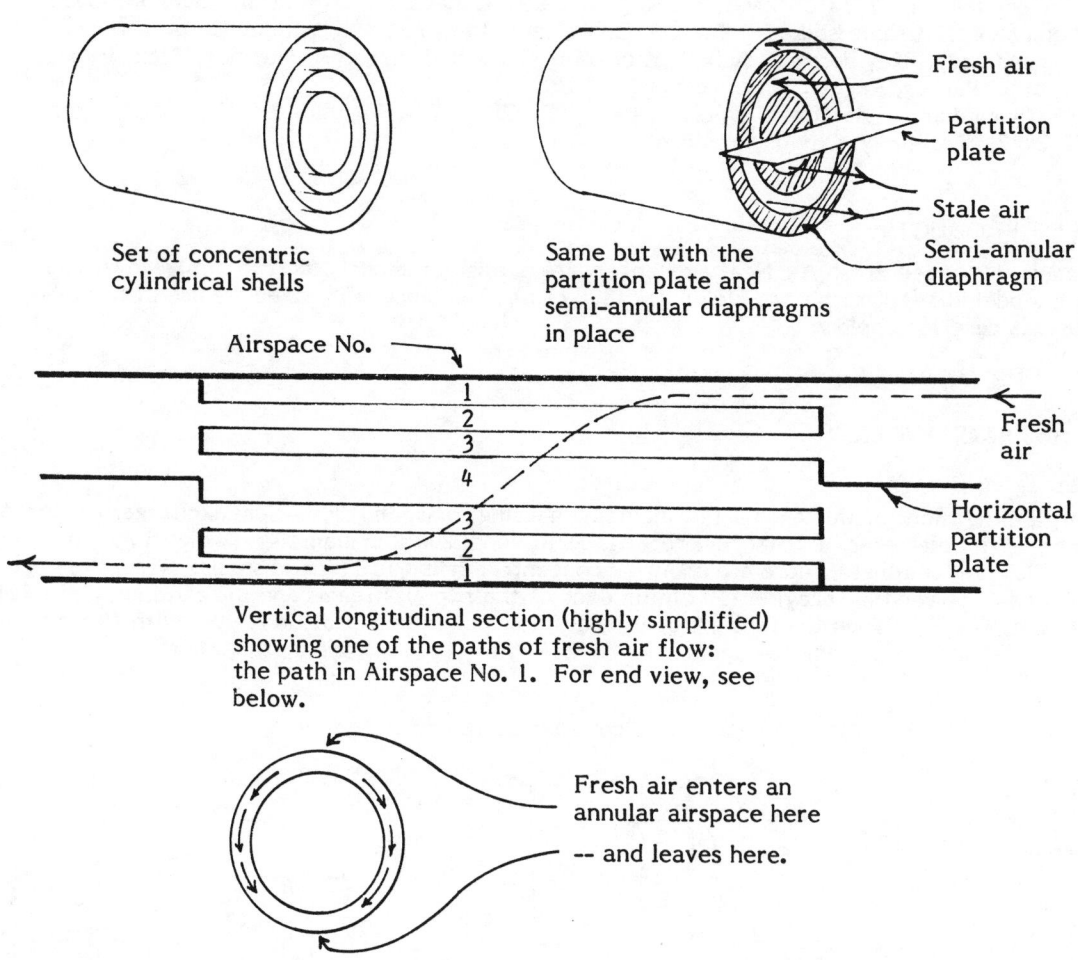

Set of concentric
cylindrical shells

Same but with the
partition plate and
semi-annular diaphragms
in place

Fresh air

Partition
plate

Stale air

Semi-annular
diaphragm

Airspace No.

Fresh
air

Horizontal
partition
plate

Vertical longitudinal section (highly simplified)
showing one of the paths of fresh air flow:
the path in Airspace No. 1. For end view, see
below.

Fresh air enters an
annular airspace here

-- and leaves here.

The following longitudinal vertical cross section is more nearly to scale.

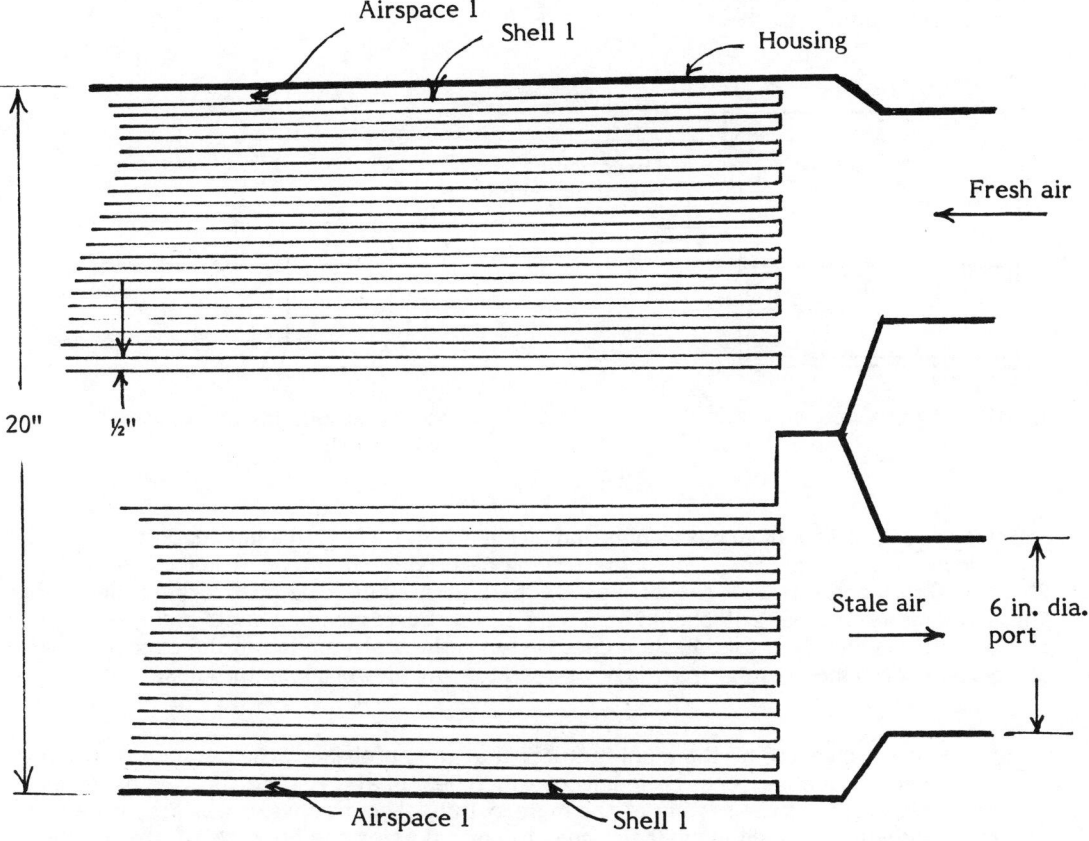

Partial longitudinal vertical cross-section
of one end of exchanger. All 22 cylindrical
shells are shown. Airspaces are ½ in. thick.

The total heat-transfer area is about 250 ft.^{2}.

There are two fans, of axial type, mounted in a common housing that is joined to the exchanger housing proper by two cylindrical sound-absorbing sleeves. Each fan has two speeds. A single "Off, Low, High" switch controls both. There are no filters.

The (8-in.-diameter) ducts extending to outdoors are insulated to R 3.5 with one inch of fiberglass. A total of 12½ ft. of insulated, 8-in.-diameter, flexible duct is supplied with the exchanger. The ducts extending to the rooms are the responsibility of the purchaser; the responsibility includes designing the duct systems and purchasing and installing the ducts.

Location and orientation: The exchanger is usually situated close to an outer wall (basement wall, crawlspace wall, or utility room wall) that is on the leeward side of the house, i.e., side away from the incident prevailing wind. The exchanger is oriented horizontally and is suspended (by special straps) close beneath the first-story floor, say.

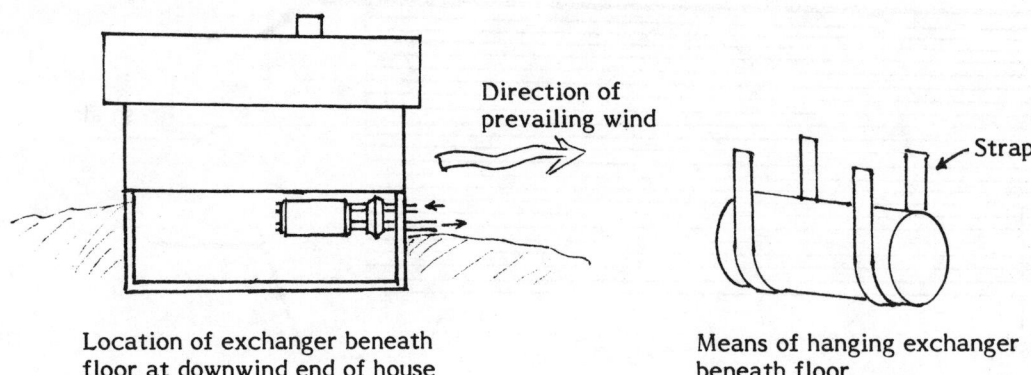

Location of exchanger beneath floor at downwind end of house

Means of hanging exchanger beneath floor

Collection of stale air: Stale air from bathrooms and kitchen flows, via grills and ducts, to a junction box and thence flows to the exchanger. The ducts serving bathrooms should be at least 5 in. in diameter (or 2½ in. x 10 in.), and the grills should typically be 4 in. x 10 in. The duct serving the kitchen should be 6 in. in diameter and the grill should be 8 in. x 10 in. Corresponding ducts and grills may be used for other rooms. No blower is used in such duct system: the stale-air fan in the exchanger suffices. Some saving of money results from lack of need for any blowers in bathrooms, kitchen, etc.

Additional controls: It is suggested by the manufacturer that the purchaser procure and install, in kitchen and utility room, manual switches by means of which he can at any time turn the exchanger fans to high speed irrespective of the setting of the main switch. The purchaser may procure and install, also, a timer that will turn such manual switches to normal after one hour. Also, the purchaser may procure and install a humidistat and connect it in such a way that whenever the indoor humidity exceeds a specified level the fans will be turned to high.

Miscellaneous components A standard-type cord and 3-prong plug (for 60-cycle, 120-v. current) are provided.

Performance: When the fans are running at high speed, the exchanger delivers, typically, 150 cfm of fresh air and ejects a like quantity of stale air. The efficiency of sensible heat recovery is 65%. Operated at low speed, which provides about 75 cfm, the exchanger has an efficiency of 79%.

Condensation: In cold weather some moisture condenses in the outgoing air paths within the exchanger. Depending on many circumstances, the amount of condensate-per-24 hour-period in an average-size house in Saskatchewan (used in typical manner by a typical family) in midwinter may be anywhere from 0.1 gal. to a few gallons. The condensate collects near the stale-air exhaust port and drips (via a small opening) into a ½-in. diameter drain hose which extends to a basement drain. A 15-ft.-long drain hose is supplied with the exchanger.

Frost formation: It is said that frosting seldom occurs, usually is of little significance, and causes no real problem unless the outdoor temperature drops to -25°F. In any event, frost build-up does no damage. In fact, the exchanger includes an integral, pressure-sensitive device that, when frosting and clogging threaten, cuts the rate of cold-air inflow to half, thus disposing of the threat. (Note: In some exchangers, when frosting occurs, the house occupant can temporarily turn off the fresh-air) blower -- which has the drawback of temporarily cutting off all inflow of fresh air via the exchanger.

<u>Summer use:</u> The exchanger is said to be of considerable use in very hot weather in summer.

<u>Price:</u> The January 1982 price to buyers in USA was $1495 FOB. Shipping the equipment 50 to 200 miles adds about $50 to the cost.

<u>Delivery period:</u> Four to six weeks after receipt of firm order accompanied by check for 30% of the total amount. Balance COD. Enercon will supply name of pertinent dealer.

<u>Warranty:</u> General one-year warranty.

<u>Maintenance:</u> Inspect motors and exchanger proper about once a year. If exchanger proper contains much dirt or dust, clean it with water, e.g., from a hose.

<u>Patent:</u> The exchanger is covered by a patent pending.

OTHER EXCHANGERS

The Model 206, Model 408, and Model 812 exchangers, providing airflow rates of 200, 400, and 800 cfm, have somewhat similar design. They employ ducts having 6-in., 8-in., and 12-in. diameters respectively. Efficiencies are about 65%. Weights: 70, 80, and 100 lb. Powers: about 100, 400, and 800 watts. Prices to U.S. buyers: $1395, $1895, and $2695. The Model 206 is suitable for use in an especially large house. All of the high-capacity models are suitable for use in large buildings or in animal housing (confinement buildings), or in indoor swimming pool areas.

Note Added in Proof

Enercon Industries Ltd. has been taken over by Blackhawk Industries, Inc., 607 Park St., Regina, Sask., Canada S4N 5N1. (306) 924-1551. David Lange, President. This company is actively producing and selling air-to-air heat-exchangers (Blackhawk Heat Recovery Systems) that are generally similar to the exchangers formerly sold by Enercon. Among the models now being sold are:

Model 15 Provides a flow of 80 cfm when connected to a duct that has 0.1 inch of back pressure. Retail price in US: $995 US.

Model 25 Provides a flow of 150 cfm when connected to a duct that has 0.1 inch of back pressure. Retail price in US: $1250 US.

Chapter 24

CONSERVATION ENERGY SYSTEMS INC. EXCHANGERS (VanEE)

The company

Range of products

VanEE R-200 exchanger

VanEE P700S exchanger

THE COMPANY

Conservation Energy Systems Inc., Box 8280, Saskatoon, Sask., Canada S7K 6C6. Tel.: (306) 665-6030. Olmstead, Rick. President. The company is the wholesale distributor of VanEE exchangers made by D.C. Heat Exchangers Ltd.

Regional Representatives

Ark Solar Products Ltd.
2676 Quadra Street
VICTORIA, B. C.
V8T 4E4 (Gil Parker)
PH. (604) 386-7643

Insulation City
4357 Canada Way
BURNABY, B. C.
V5G 1J3 (Len Fleming)
PH. (604) 430-2292

Tri-Energy Tech. Inc.
1540D Highway 97 South
KELOWNA, B. C.
V8T 4E4 (Ken Farrish)
PH. (604) 769-3080

Encotech Energy Systems Ltd.
10062 - 80th Avenue
EDMONTON, Alberta
T6E 1T2 (Chris Bamford)
PH. (403) 432-1086

Fireside Conserver Products
835 C Broadway Avenue
SASKATOON, Saskatchewan
S7N 1B5 (David Van Vliet)
PH. (306) 665-6707

P & G Plumbing
Box 778
NIPAWIN, Saskatchewan
S0E 1E0 (Peter Gilbert)
PH. (306) 862-4819

Lay Design & Fabricating
EDEN MILLS, Ontario
N0B 1P0 (Richard Lay)
PH. (519) 821-8478

Alternate Heating Ltd.
621 Rothesay Avenue
SAINT JOHN, New Brunswick
E2H 2G9 (Donald Hemmings)
PH. (506) 696-2321

Centennial Plumbing & Heating Ltd.
212 Avenue B South
SASKATOON, Saskatchewan
S7M 1M4 (Heinz Beckman)
(306) 244-8211

Home Energy Audit Ltd.
653 George Street
SYDNEY, Nova Scotia
B1P 1L2 (Bill McDonald)
PH. (902) 539-5095

Sunliner Developments Ltd.
2618 South Province Drive
LETHBRIDGE, Alberta
T1K 0J4 (Isabelle Hamilton)
PH. (403) 327-3570

RANGE OF PRODUCTS

VanEE R-200   200 cfm. 48 x 15 x 22 in. $795 Canadian. Most popular model; for general use; mounted beneath ceiling.

VanEE P700S   700 cfm. Large size. For swimming pool enclosures. Purchaser must provide blowers, ducts, and some controls. $795 Canadian.

VanEE R200 EXCHANGER

Summary

The VanEE R200, double-crossflow, ceiling-mounted exchanger is 48 in. long, 15 in. wide, and 22 in. high. Operated at 200 cfm, it has an efficiency of sensible-heat recovery of about 71%. Blower speed may be varied locally or remotely. A dehumidistat is included. Price: $795 Canadian or $695 US.

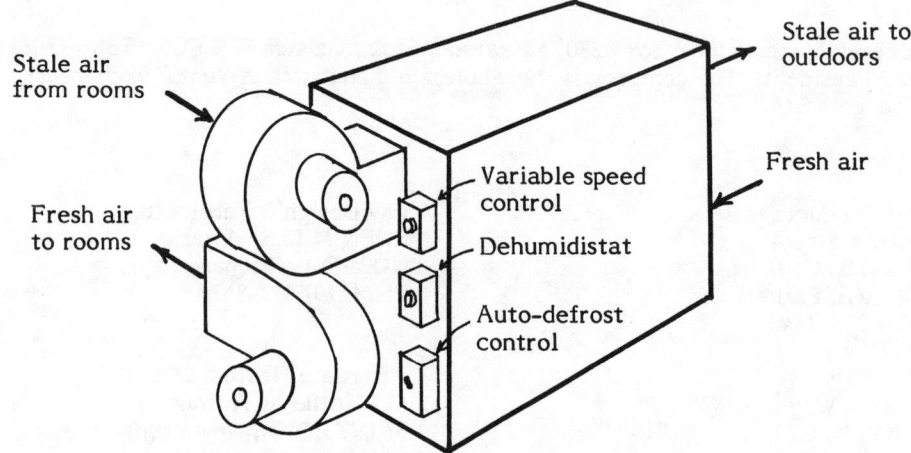

The VanEE R200 exchanger. Perspective view.

Exchanger Proper

This is of double crossflow type. The heat-transfer sheets are fluted, are of 0.012-in.-thick polypropylene, and are spaced 0.2 in. apart. The housing is of painted galvanized steel insulated internally with foam to prevent moisture condensation on outside.

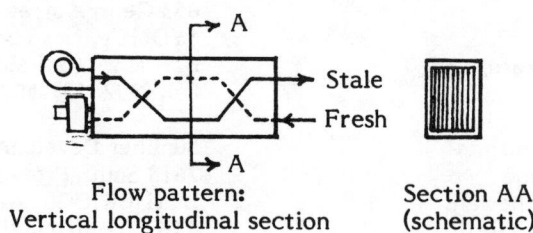

Flow pattern:
Vertical longitudinal section

Section AA
(schematic)

Blowers: There are two centrifugal-type blowers. Each provides 250 cfm of airflow before the ducts are attached and 200 cfm after the ducts have been attached (if the pressure drop in the ducts is 0.33 in. water). Each blower has its own 1.2 amp. motor. The fresh-air blower is automatically controlled by a temperature sensor which turns off this blower whenever the fresh air that is leaving the exchanger and entering the room is below a certain temperature and turns the blower on again when the entering air reaches a certain higher temperature.

Ducts: The necessary ducts, 6 in. in diameter, are supplied by the purchaser. The duct for incoming fresh air should be as short as feasible and should be insulated externally with fiberglass, and the insulation should be covered by a high-quality vapor barrier that is well sealed.

Filters: None. If the indoor air contains much dust (e.g., air from a clothes dryer) the purchaser should obtain and install a filter in the stale air stream, upstream from the exchanger proper. Specific recommendations for the filter are available from the exchanger supplier.

Controls: Blower speeds are of manually continuously variable type. A Honeywell dehumidistat is included, and whenever the indoor relative humidity exceeds a specified value, for example 38%, the dehumidistat overrides other controls and causes the exchanger to operate at highest speed. Manual switches may be provided in kitchen, bathroom, etc., and these too can override the normal running regimen. The dehumidistat can be located at the exchanger or may be remotely located, e.g., in some other room.

Cord and plug: None. To conform to code, the installer should arrange to have the exchanger directly wired by a qualified electrician. Also, a main disconnect switch must be installed, to permit servicing the exchanger.

Pilot light and fuse: None. The exchanger should be wired to a separate circuit breaker.

Disposal of condensate: A 10-ft-long, ½-in.-diameter drain hose of vinyl plastic is provided.

Defrosting: Defrosting, when necessary, is accomplished automatically. The fresh-air-intake blower is turned off automatically, for a short period; this permits the outgoing air to remain warm and to melt the frost. The automatic-control sensor is a temperature sensor; it turns off the fresh-air blower whenever the temperature of the fresh air that is leaving the exchanger and entering the room is below a certain temperature, and turns this blower on again when the temperature in question rises to a certain temperature. Typically, the two temperatures chosen are $35^{\circ}F$ and $44^{\circ}F$ respectively. By turning a small wheel (which becomes accessible when a cover plate is removed) the house occupant can vary the temperature settings.

Performance: When a typical set of ducts has been installed and the flowrate in each airstream is about 200 cfm, the efficiency of sensible-heat recovery is 71%. When the flowrate is halved, the efficiency increases to 83%.

Installation: The exchanger is oriented horizontally and is suspended by four straps (supplied with the exchanger) beneath the ceiling of basement or utility room. Typically, stale air is collected from kitchen, bathroom, etc., and fresh air is discharged at a single location.

Price, etc.: $795 Canadian F.O.B. or about $695 in US dollars. Delivery period: 3 or 4 weeks. A manual is provided. Warranty covers defects in materials and workmanship for one year and covers the exchanger proper for five years. Note: US customs duty is about 6%.

Maintenance: If the exchanger proper should become clogged with dust, the house occupant should detach the exchanger, remove the two blowers, stand the assembly on end, and flush out the dust with water. A design modification that makes the exchanger proper more accessible may be put into effect in 1982.

VanEE P700S EXCHANGER

This exchanger, designed for use in swimming pool enclosures or other very humid spaces, is capable of providing a very high (700 cfm) flowrate.

The complete package includes little more than the exchanger proper. The purchaser himself must supply the blowers and any special controls. An auto-defrost control is provided with the exchanger. Dimensions of exchanger: 36 in. x 36 in. x 22 in. The case, of galvanized steel, is insulated. The airflow apertures to which ducts are to be connected are 12 in. x 12 in. Retail price: about $795 Canadian F.O.B., or $695 in US dollars.

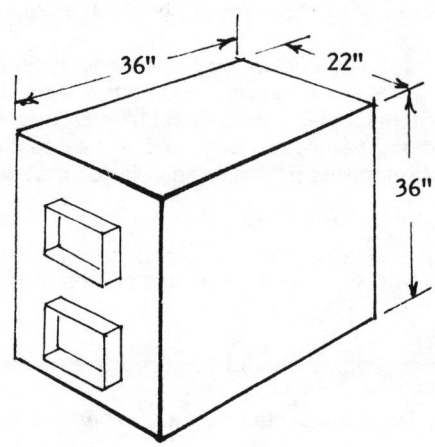

Chapter 25

TEMOVEX AB EXCHANGERS

THE COMPANY

Temovex AB, Box 111, S-265 01 Åstorp, Sweden. (Bangatan 5, Åstorp is the "visiting address".)
Tel.: 042 550 20. Templer, Börge, is Manager. US distributor: none.

THE 480-S EXCHANGER

This is a parallel-plate, countercurrent-flow, air-to-air heat-exchanger designed to serve several
rooms. Dimensions: 65 in. x 14½ in. x 17.7 in. high. At one end there are two centrifugal blowers.
Filters are situated upstream from the exchanger proper. At each end of the housing there are two
160-mm ports. There is a small drain tube at one end. The exchanger is mounted horizontally. There
is 2 in. of external insulation. Overall weight: 120 lb. (Exchanger proper: 57 lb.). The control switch,
usually installed in kitchen, provides choice of several rates of airflow. Normally 50-cycle electric
power is assumed, but alterations to accommodate 60 cycle power can be provided. Typically the
exchanger is mounted above the ceiling and stale air is vented through the roof. With 150 cfm airflow the efficiency of sensible heat recovery is 79% and power consumption is 190 w.; with 75 cfm
airflow, efficiency is 82% and power consumption is 95 w. March 1982 price: About $1000.

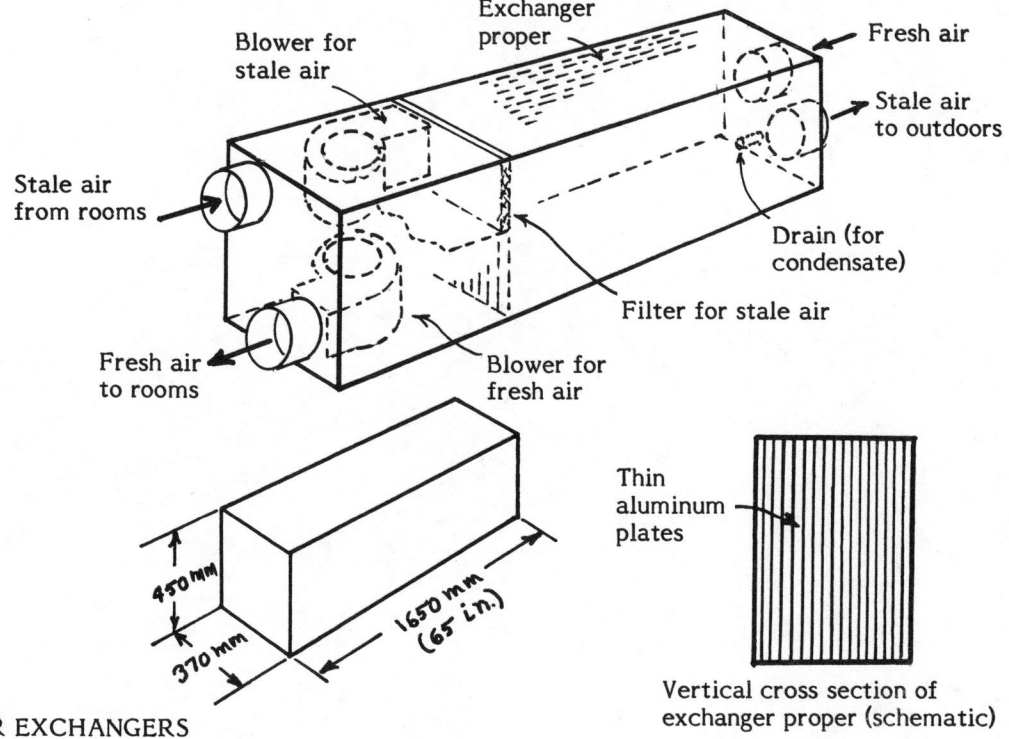

Vertical cross section of
exchanger proper (schematic)

OTHER EXCHANGERS

VX Exchanger (Cabinet-type)	New model, of upright type for installation in laundry room. Dimensions: 17 in. x 24 in. x 83 in. high. Weight: 220 lb. Price: $1100. Can be joined to a "VU" heating device.
1500 Exchanger	Provides very high flowrates: rates as high as 2000 m^3/hr.

Chapter 26

AIR CHANGER CO. LTD. (THE) AIR CHANGER

The company
Range of products
Deluxe Air Changer
Other exchangers

THE COMPANY

The Air Changer Co., Ltd., 334 King St. East, Toronto, Ont. Canada M5A 1K8. Tel.: (416) 863-1792. Elizabeth White, President. Dara Bowser, Sales Manager.
   Developer of exchangers: Allen-Drerup-White, Ltd. Same address as above.

RANGE OF PRODUCTS

The main product is the Deluxe Air Changer, but there are also simpler, less expensive models called Basic and Standard.

DELUXE AIR CHANGER

This counterflow air-to-air heat-exchanger is 54 in. long, 22 in. deep, and 14 in. wide. It provides air-flows up to 150 cfm, with a pressure drop of 0.1 to 0.2 in. water. Two axial-type fans, each requiring 37 watts, are employed. At each end of the housing there are two 6-in.-diameter openings with 6-in.-diameter collars for duct connections. Shipping weight: 85 lb.

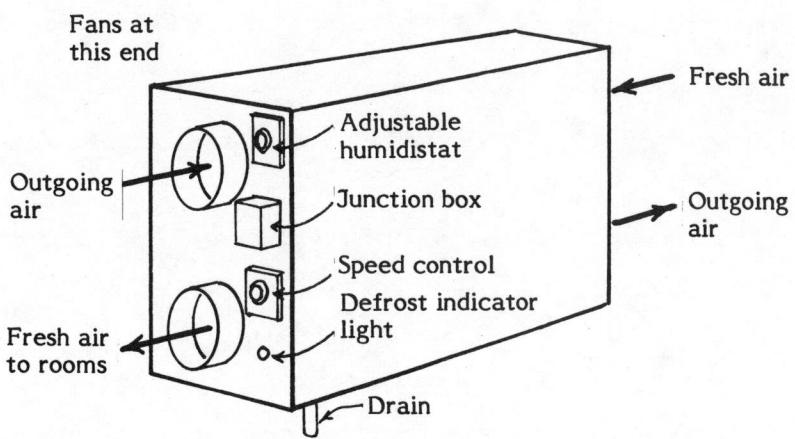

The exchanger proper consists of twinwall Coroplast sheets -- plastic sheets that contain a great many slender passages for air. One airflow utilizes the slender passages and the other utilizes airspaces left between twinwall sheets. The total heat-transfer area is 300 ft.^{2} The materials are not permeable to water.

Operated at high speed, each fan provides a flowrate of 150 cfm. Operated at "manual control", any speed from 0 to 150 cfm may be provided. In each room where stale air is collected by a duct there is an overriding manual control. Automatic switch to high-speed operation occurs also when the dehumidistat, situated within the exchanger proper, senses humidity above a preselected limit.

Condensate collects at the low end of the exchanger and leaves via a drain.

Efficiency, with 150, 100, and 50 cfm flowrates: 85%, 89%, and 92% respectively.

The exchanger, in its internally insulated, 22-gauge galvanized steel housing, is mounted beneath the floor, between joists. It is supported by steel straps.

Ducts and grills, not supplied with the equipment, must be provided by the purchaser. Main and subsidiary ducts should be 6, 5, or 4 in. in diameter. The combined area of the stale-air intake ducts in the rooms should be at least 80 in.^{2}.

If frost begins to clog the exchanger passages for outgoing air, a within-exchanger temperature sensor responds to the abnormally low temperature of the incoming air entering the room and turns off the blower serving the incoming airstream -- until the temperature in question returns to normal.

Maintenance: If, after a year or several years, the flowrates appear to have decreased, the house occupant should suspect that the exchanger passages have become partially clogged with dust. He should disconnect the exchanger's ducts, wiring, and drain hose, detach the suspension straps, lower the exchanger to the floor, detach and remove end-plate at warm end of the exchanger, and clean the outgoing-air passages of the exchanger proper with soapy water.

Price: $860 U.S., with one-year general warranty and 5-year warranty on core.

OTHER EXCHANGERS

Model Called Standard

As above except that each of the two blowers provides an airflow of 100 cfm and the only control provided is the defrost control. $670 U.S.

Model Called Basic

This includes just the exchanger proper and the housing. No blowers or controls are included. $400 U.S.

Chapter 27

BESANT-DUMONT-VAN EE EXCHANGER

This exchanger, invented a few years ago by R. W. Besant, R. S. Dumont, and D. Van Ee of the Dept. of Mechanical Engineering of the University of Saskatchewan, has been described in detail in their report "An Air to Air Heat Exchanger for Residences" (Bibl. B-251a).

The heart of this counterflow, vertical-parallel-plate exchanger is a set (stack) of 0.006-in.-thick sheets of polyethylene -- or, more exactly, a single very long sheet that has been folded back and

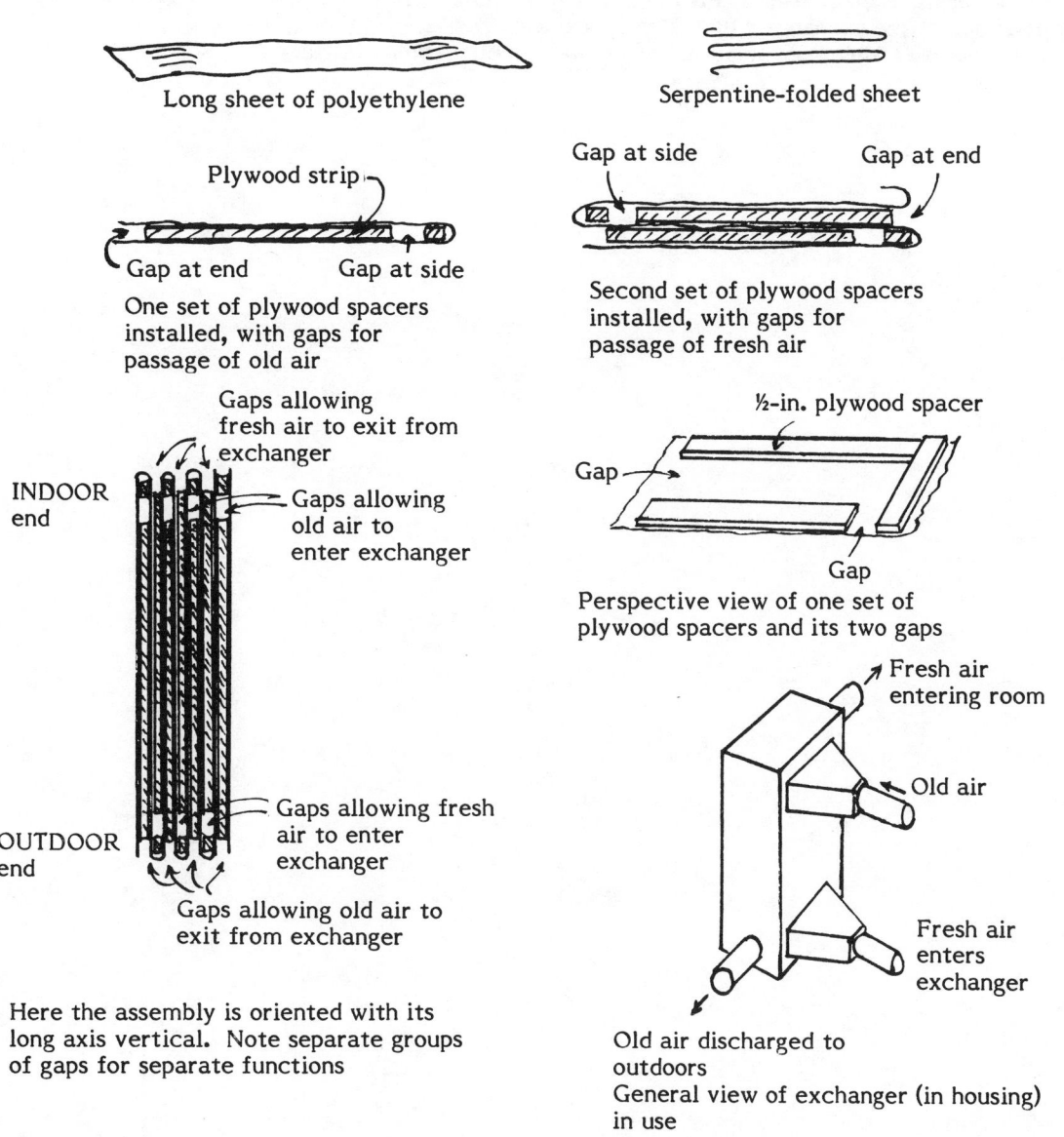

Long sheet of polyethylene

Serpentine-folded sheet

One set of plywood spacers installed, with gaps for passage of old air

Gap at end   Gap at side
Plywood strip

Gap at side   Gap at end

Second set of plywood spacers installed, with gaps for passage of fresh air

Gaps allowing fresh air to exit from exchanger

INDOOR end

Gaps allowing old air to enter exchanger

½-in. plywood spacer

Gap

Gap

Perspective view of one set of plywood spacers and its two gaps

Gaps allowing fresh air to enter exchanger

OUTDOOR end

Gaps allowing old air to exit from exchanger

Here the assembly is oriented with its long axis vertical. Note separate groups of gaps for separate functions

Fresh air entering room

Old air

Fresh air enters exchanger

Old air discharged to outdoors
General view of exchanger (in housing) in use

150

forth many times in serpentine manner. There are 36 folds and 37 flat areas. Between successive flat areas are ½-in.-thick spacers, or frames, of plywood. Each such frame has two gaps, or ports, to let air in and out. The 36 airspaces between plastic-sheet flat areas constitute two sets: old-air-out set and fresh-air-in set. The former exploys airspaces 1, 3, 5, etc., and the latter employs airspaces 2, 4, 6, etc. The assembly, which is 65 in. x 24 in. x 19 in., is mounted (just inside the house) with the long axis vertical. The old air travels downward through it and the fresh air travels upward through it. Because the total area of plastic (the heat-exchange area) is large, and because the plastic is thin, the heat-exchanger efficiency may be fairly high, such as 75% when the rate of airflow is 100 cfm and 85% or 90% when the airflow is 50 cfm. (I am told that somewhat higher efficiency may be attainable if the exchanger is made with great care and if, in particular, the plastic sheets are kept so tight that no "ballooning" can occur.) Airflows are maintained by two small (20-w) blowers, one for each airstream.

In mid-winter as much as 25 lb. of ice may form, each 24-hr. period, in the lower part of the exchanger. Defrosting procedures, described in B-251a, must be used.

I understand that there are some advantages in substituting sheets of 28-gauge galvanized steel for the plastic sheets. The sheets remain strictly flat. Also they are not subject to damage during assembly of the exchanger; that is, they are less fragile.

Also sheets of aluminum perform better than sheets of thin plastic.

Chapter 28

OTHER EXCHANGERS

Advanced Idea Mechanics Ltd. exchangers

Aldes Riehs U.S.A. exchanger

Automated Controls & Systems exchanger

Bahco Ventilation Ltd. exchangers

Cargocaire Engineering Corp. exchangers

D. C. Heat Exchangers Ltd. exchangers

Del-Air Systems Ltd. exchanger

Dravo Corp. exchangers

Ener-Corp Management Ltd. exchanger

Flakt exchanger

Gaylord Industries exchangers

Genvex Energiteknik A/S exchangers

Heatex AB exchanger

Memphremagog Group exchangers

Munters AB exchanger

Norlett exchanger

Passive Solar Products Ltd. exchanger

Temp-X-Changer Division exchanger

Wing Co. (the) exchanger

Drainage-pipe-type exchanger

Saunders exchanger

ADVANCED IDEA MECHANICS LTD. EXCHANGER

Address: 26 Ski Valley Crescent, London, Ontario, Canada N6K 3H3. Tel.: (519) 471-5573.
Henderson, Robert: President.

This company, with some support from Canada Mortgage and Housing Corp., has been developing a Model AIM-120 crossflow, 16 in. x 16 in. x 22 in., 55 lb., air-to-air heat-exchanger employing aluminum heat-transfer sheets, two squirrel-cage blowers (driven by a single motor, to insure keeping the two blowers at the same rpm), filters, and a built-in 500-w preheater to eliminate the threat of frost formation. Flowrate: about 100 cfm if duct pressure is 0.3 to 0.4 in. water. Core and filters are easily removable for cleaning. The exchanger is expected to be ready for sale early in 1982, at a retail price of $650 Canadian or $550 US F.O.B.

The company also sells Mitsubishi Lossnay exchangers.

ALDES RIEHS U.S.A. EXCHANGER

Address: 157 Glenfield Rd., Sewickley, PA 15143. Tel.: (412) 741-2659.

This firm is the US importer of the Aldes VMPI Exchanger made by Aldes Co. of Lyon, France. In October 1981 arrangements for import were being completed. Model installations have been completed in Pittsburgh, PA and near Frederick, MD.

The exchanger is mainly of counterflow type but partly (near the ends of the exchanger chamber) of crossflow type. The heat-transfer sheets are of plastic and are amply rigid. The total area for heat-transfer is 210 ft.^{2}. The housing is of sheet metal and is insulated. The dimensions of the exchanger, per a rough estimate, are 57 in. long by 18 in. high by 11 in. deep. The weight is about 50 lb. An electric preheater comes into play when the outdoor temperature is very low. Blowers are not included, they must be provided by the purchaser. The price of the exchanger is $750 F.O.B.

According to test results reported by investigators at Lawrence Berkeley Laboratory (F-70), an incomplete exchanger (lacking flexible ducting, diffusers, etc.) was found to have high efficiency of sensible-heat-transfer: 73% with a 60 cfm flowrate and 68% with 230 cfm flowrate. The airstream static pressure drop was 0.05 in. water with 60 cfm flowrate and 0.4 in. water with 230 cfm flowrate.

English branch -- Aldes, PO Box 3, Brookside Industrial Estate Rustington, Littlehampton, Sussex, BN 16 3LH, England -- has distributed brochures in English.

AUTOMATED CONTROLS & SYSTEMS EXCHANGER

Address: 500 East Higgins Rd., Elkgrove, IL 60007. Tel.: (312) 860-6860.
The company sells cross-flow exchanger cores of many types.

BAHCO VENTILATION LIMITED EXCHANGERS

Address: Bahco House, Beaumont Rd., Oxfordshire, England OX16 7TB. Distributes equipment made in Sweden.

The Minimaster exchanger, to be mounted above the kitchen cooking stove, is of cross-flow type and weighs 90 lb. The exchanger proper contains many parallel aluminum plates and is served by two centrifugal type blowers each requiring 85 w. A manually operated switch permits choice of any of three speeds or <u>off</u>. A manually operated damper controls the proportions of air taken from the above-stove hood and from other rooms (living room, bathroom, etc.). Each airstream is served by a filter situated upstream from the exchanger proper. A thermostatically controlled 1 kW electric heater is situated in the stream of fresh air emerging from the exchanger; the heater may be turned on or off manually. With flowrates of about 100 cfm the efficiency of sensible-heat recovery is about 60 or 65%. There are two models: one for average-size house (1400 ft^2) and the other for larger house (1700 ft^2).

The parent company (manufacturing company) is AB Bahco Ventilation, (S-) 199 81 Enköping, Sweden. The export manager is Björn Möller.

The Minimaster exchanger includes 185 heat-exchange plates. Each is of aluminum and is 200 mm x 200 mm x 0.3 mm; the plates are 2 mm apart. A 13-page "Installation Instructions" manual provides detailed information on ducts layout, insulation of ducts, grills, etc.

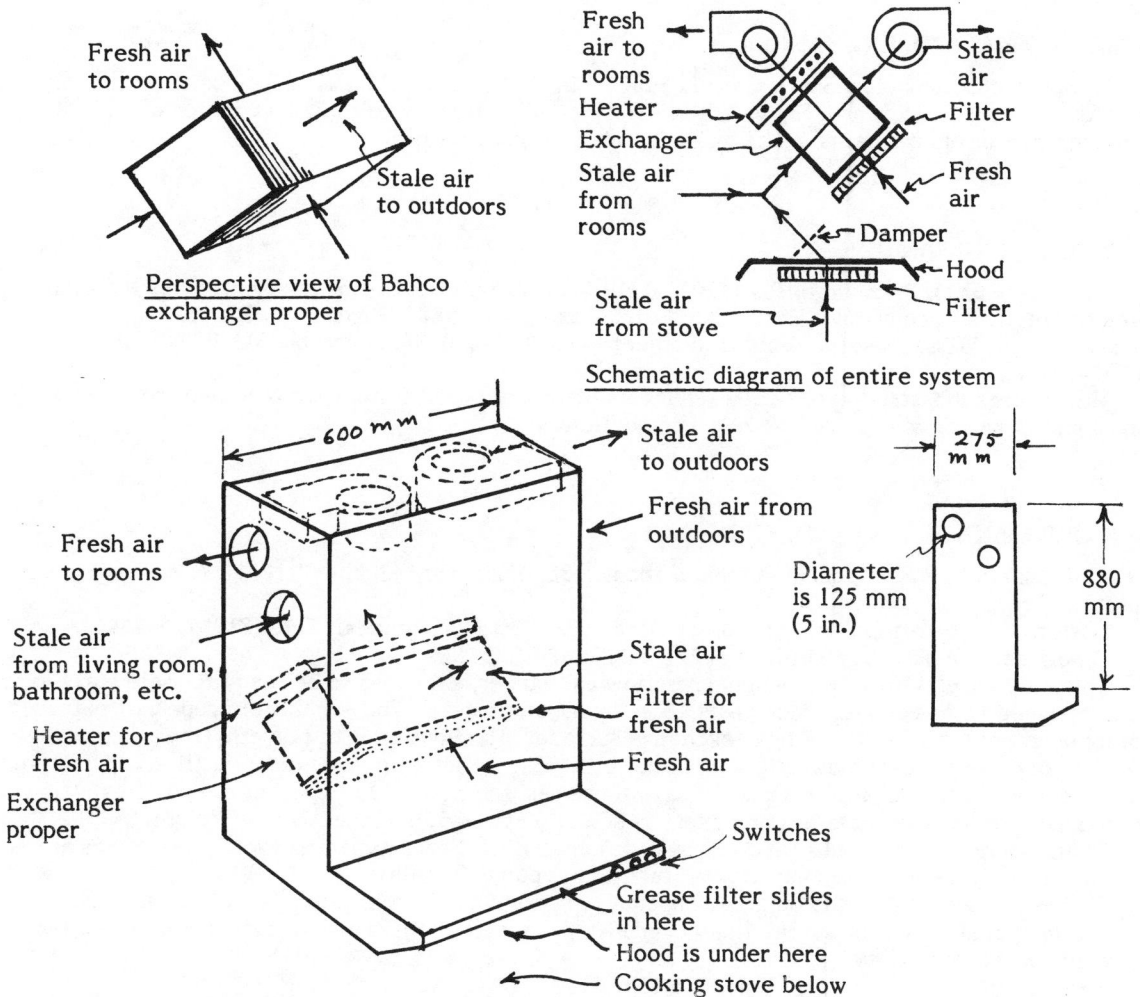

Perspective view of Bahco exchanger proper

Schematic diagram of entire system

Incomplete X-ray view of Bahco exchanger installed above cooking stove. (Omitted from drawing are baffles, damper, thermostat, drain connection, pilot light, etc.)

CARGOCAIRE ENGINEERING CORP. EXCHANGERS

Address: 79 Monroe St., Amesbury, MA 01913. Tel.: (617) 388-0600.

Makes rotary and also fixed-type air-to-air heat-exchangers for industrial use etc. The smallest of the rotary exchangers employs a 36-in.-diameter rotor and supplies 3000 cfm. The rotors are of fiberglass and plastic, or of aluminum. Some transfer sensible heat only and others, desiccant impregnated, transfer sensible heat and latent heat. The smallest fixed-type exchanger supplies 100 cfm and is priced at about $2200. The company obtains many of its exchanger cores from A. B. Munters, P.O. Box 7093, S-191 07 Sollentona, Sweden.

D. C. HEAT EXCHANGERS LTD. EXCHANGER

Address: Box 8339, Saskatoon, Sask., Canada S7K 6C6. Van Ee, Dick. President.

Makes Van Ee exchangers for houses and Del-Air exchangers for animal barns. These two classes of exchangers are sold at wholesale by Conservation Energy Systems Inc. (see Chap. 24 for detailed account) and by Del-Air Systems Ltd. respectively.

DEL-AIR SYSTEMS LTD. EXCHANGER

Address: Box 2500, Humboldt, Sask., Canada SOK 2AO. Tel.: (306) 682-5011. Key persons: Benning, Larry. Also Novecosky, Edward, Sec'y.-Treas. This company is the wholesale distributor of the Del-Air exchangers (for animal barns) made by D. C. Heat Exchangers Ltd.

DRAVO CORP. EXCHANGERS

Address: PO Box 9305, Pittsburgh, PA 15225. Tel.: (412) 777-5932. Dormetta, Ron. Sales Manager. Related or antecedent company: Hastings Industries of PO Box 9358, Pittsburgh, PA 15225. Representative in Massachusetts: Long Equipment Co., PO Box 0108, Holbrook, MA 02343. Tel.: (617) 767-3595.

Makes large industrial-type heat-exchangers of rotary type. Rotors are obtained from Berner International Corp. Smallest unit has 2000 cfm airflow.

ENER-CORP MANAGEMENT LTD. EXCHANGER

Address: 2 Donald St., Winnipeg, Manitoba, Canada R3L OK5. Tel.: (204) 477-1283. A key engineer: Giesbrecht, Peter.

Eastern Canada office: 440 St. Joseph Blvd., Hull/Ottawa, Canada. Tel.: (819) 777-4446.

US office: 359 Rt. 111, Smithtown, NY 11787. Tel.: (516) 724-2111.

This 70 in. x 13.5 in. x 16-in.-high, crossflow exchanger, expected to be ready for sale early in 1982, is designed to fit between floor joists 16 in. apart on centers. The exchanger proper, of extruded polypropylene, provides 300 ft^2 of heat-exchange surface. The housing is of polyethylene. The two 120-v., 1.8-amp. centrifugal blowers are at either end. Each airstream is served by a filter. Preliminary measurements of heat-exchange efficiency gave these results: with 130 cfm, 95 cfm, and 75 cfm flowrates, the efficiency of sensible-heat recovery is 80%, 86%, and 95% respectively. Weight: 80 lb.

Operation is initiated automatically by a dehumidistat. Several manual override switches are provided; they may be installed in areas that tend to have high humidity.

As the temperature of the outgoing air emerging from the exchanger falls and becomes close to 32°F, a temperature sensor automatically reduces the speed of the fresh-air blower sufficiently to avoid risk of frost formation.

The bottom panel of the housing can be removed to permit removing and cleaning the exchanger proper.

Price: To be in neighborhood of $900 Canadian. Distribution in USA will be through an office in Indiana.

FLAKT EXCHANGERS

Flakt exchangers are made by the well-known Swedish company Svenska Flakt Fabriken. Until recently the marketing of small Flakt exchangers, such as the Rexovent, was handled by a subsidiary located in Fort Lauderdale, Florida, and in Winston-Salem, North Carolina. But by January 1982 such marketing had been terminated. Accordingly the following paragraphs are of diminished interest.

The Rexovent exchanger is similar to the Genvex exchanger. Airflow is mainly of crossflow type. The exchanger proper is said to include fins, or, more exactly, thin sheets of aluminum that have been accordion or zig-zag-folded so as to convert the large simple airspaces between plates into myriad slender spaces that are triangular in cross section. Besides helping heat-transfer (and keeping the airflow laminar?), the fins help maintain uniform spaces between plates. Total heat-exchange area: 84 ft^2. Total weight: 80 lb.

The exchanger includes an electrical heating element which (1) insures that the fresh air delivered to the rooms is never colder than 52°F and (2) prevents formation of frost or ice. The heating element presumably slightly reduces the efficiency of heat-recovery from the outgoing air.

According to test results reported by Lawrence Berkeley Laboratory (F-70), the exchanger has an efficiency of sensible-heat-recovery of 68% with a 60 cfm airflow and 56% with a 230 cfm airflow. Airstream static pressure drop with flowrates of 60 cfm and 230 cfm was 0.1 in. water and 1.0 in. water respectively. Total electrical power used by the blowers (with flowrates of 60 to 170 cfm) was 140 to 160 watt. (Refs.: F-70, F-700).

The company makes rotary exchangers also, per R-25.

GAYLORD INDUSTRIES EXCHANGERS

Address: PO Box 558, 9600 SW Seely Ave., Wilsonville, OR 97070. Tel.: (503) 682-3801. Black, David K. Vice President.

The company makes and sells ventilators and heat-recovery equipment for commercial cooking installations, especially for large stoves in commercial and institutional kitchens.

Ventilator

This is a versatile, high-performance hood, situated above the stove, and a roof-mounted blower. Within the hood there are baffles that cause sharp direction-changes in the stream of outgoing air; the centrifugal forces associated with the direction changes tend to throw grease particles out of the airstream and into special grease-collecting gutters. Efficiency of grease-particle capture: 95%.

Whenever the blower is shut off (for example, at the end of the day), a cleaning cycle starts automatically. Hot detergent water frees and captures the grease and carries it to a drain.

An automatic fire-thwarting feature is included. When and if a thermostat senses that the airstream temperature reaches 350°F, it causes a baffle to close, stopping the flow of air from hood proper to duct system. Also, it turns off the blower. Also it initiates a spray of water into the interior of the ventilator.

A wide variety of ventilator sizes and shapes are available. Much use is made of 18-gauge stainless steel (No. 4 finish). Typical airflow rate: 250 cfm per foot of length of hood. Pressure drop: 1.3 to 1.5 in. of water.

Typical specifications of ventilator: 3 to 30 ft. length, 500 to 3000 lb. wgt.

Heat-Exchanger

This consists of an array of horizontal, parallel, heat-pipes made by Q-Dot Corp. The two airstreams (fresh air, stale air) pass through the array side-by-side, separated by a partition that insures that the fresh air and stale air cannot become mixed. Each airstream is served by a duct and a blower. Typically, the volume of flow in the fresh air stream is 80% of that in the stale air stream. The system as a whole can be provided with an auxiliary heater that can insure that the air entering the building is warm enough. The heat-exchanger may be mounted on the roof or in a utility room or mechanical room. The efficiency of sensible-heat recovery is said to be about 55 to 65%, typically, depending on exact design and operating mode. See company's 11-p. brochure "HRU Heat Reclaim Performance". See also Chapter 21 on Q-Dot exchangers.

Cost

About $2.10 per unit of total flow (sum of flows in two airstreams). Somewhat more (per unit of flow) for lower flowrates. Example: If the individual flowrates are 5000 and 4000 cfm, the total is 9000 cfm and the cost, F.O.B., of the exchanger is about $20,000. The ventilator costs about $12,000, making a total of about $32,000 for exchanger and ventilator.

GENVEX ENERGITEKNIK A/S EXCHANGERS

Address: Blegdamsvej 104, DK-2100, København Ø, Denmark.

The exchanger is of crossflow type and contains 93 ft^2 of heat-exchange sheets of aluminum. Two fans (or blowers) are used; each is driven by a 220-volt, single phase, motor designed for use with 50-cycle current but capable of running on 60-cycle current. There are two filters. The housing is of steel and is insulated; one side of the housing may be removed to give access to the exchanger proper, fans, and filters. Total weight: 150 lb.

According to test results reported by Lawrence Berkeley Laboratory (F-70), an imperfectly assembled exchanger of this type was found to have an efficiency of 64% when the flowrate was 60 cfm and an efficiency of 46% when the flowrate was 230 cfm. Total electrical power usage (with flowrates of 70 to 160 cfm) was 130 to 150 watt.

HEATEX AB EXCHANGER

Address: Hammarsvagen, Sweden. Makes exchanger cores.

MEMPHREMAGOG GROUP EXCHANGERS

Address: PO Box 456, Newport, VT 05855. Tel.: (802) 334-8821. Some key persons: Mr. Armand Lepage, Mr. Blair Hamilton.

For a year or more this group, with some support from Dept. of Energy and Department of Agriculture, has been developing three kinds of air-to-air heat-exchangers. Only one of these, called exchanger for house, is now in routine production.

Exchanger For House

This exchanger employs counterflowing, laminar-flow airsteams. Dimensions: 50 in. x 28 in. x 14 in. The exchanger proper consists mainly of twinwall sheets of Coroplast, a 90-10 combination of polypropylene and polyethylene. Each of these extruded sheets is 5.3 mm thick and has partitions every 5 mm. Spacers along the edges keep the sheets 5 mm apart. Weight: 85 lb. Fresh air flows along the slender (5 x 5 mm) channels, and stale air flows (in the opposite direction) in the clear spaces between sheets. The fresh air enters at one end of the assembly, whereas the stale air enters from the side, near one end; thus -- near the ends -- the flows are to some extent crossflows. Flowrate in each stream is about 100 to 150 cfm and the efficiency of sensible-heat recovery is about 80%.

5.3 mm

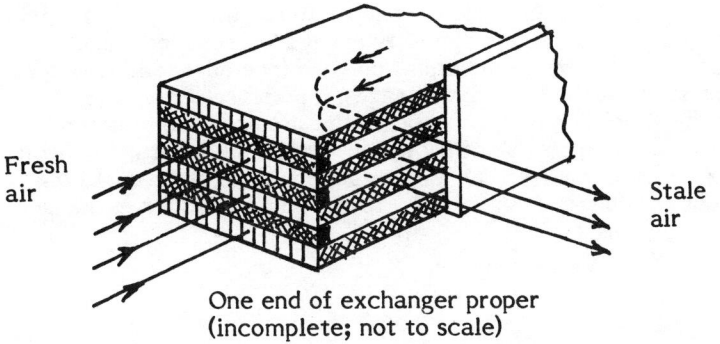

Fresh air →

Stale air →

One end of exchanger proper
(incomplete; not to scale)

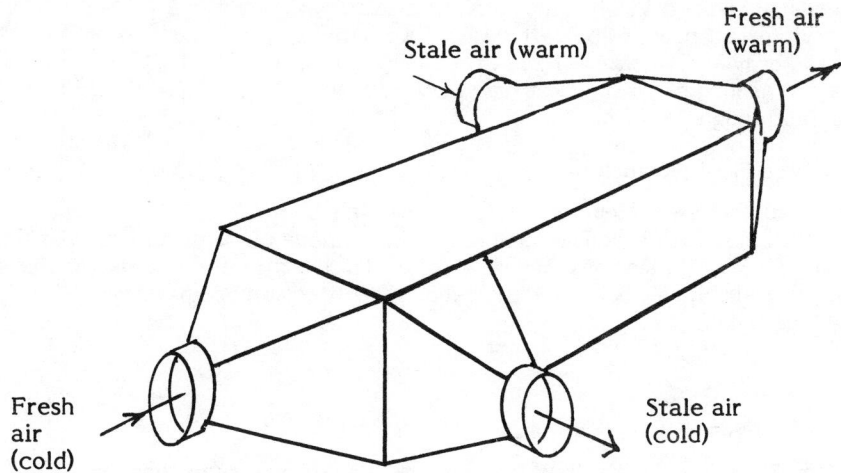

Stale air (warm)

Fresh air
(warm)

Fresh air
(cold)

Stale air
(cold)

One experimental model of exchanger for house

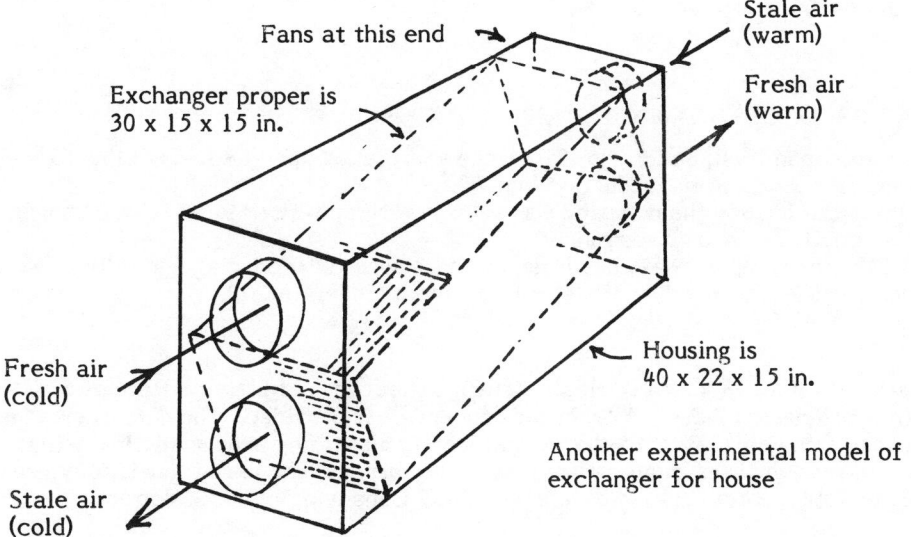

Fans at this end

Stale air
(warm)

Exchanger proper is
30 x 15 x 15 in.

Fresh air
(warm)

Fresh air
(cold)

Housing is
40 x 22 x 15 in.

Stale air
(cold)

Another experimental model of
exchanger for house

158

The two single-speed 32-watt fans can be turned on and off separately. In each bathroom there is a timer switch; turning it on (manually) causes the exchanger to run for (say) 10 minutes. A dehumidi-stat in the kitchen turns on the fans whenever the humidity exceeds a specified value. Because the exchanger is tilted, any condensate that forms collects at one end (one edge) and leaves via a drain tube. If frost forms and begins to block the stale air passages, a pressure-sensor detects the increase in pressure drop and turns off the stale-air fan until the pressure drop reverts to normal.

The exchanger is installed in basement or utility room. Ducts collect stale air from bathrooms, kitchen, and living room. Maintenance: if exchanger proper accumulates much dust etc., remove it, take it outdoors, and wash it out with water from a hose. Price: about $500 for exchanger proper, or $800 complete with fans and controls.

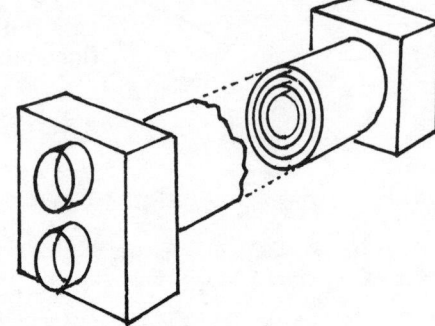

Exchanger For Greenhouse

This exchanger, still under development, employs a set of concentric cylindrical tubes of polyethylene. Two axial fans are used and provide flowrates of 100 cfm. The exchanger may be as long as 25 ft. Cost may be very low -- in the $200 to $300 range.

Exchanger For Shed For Livestock

This exchanger, still under development, emphasizes high flowrate: 500 cfm. Cross flow is used, and much use is made of twinwall Coroplast sheets such as are discussed above. Overall dimensions may be 4 ft. x 4 ft. x 1 ft., with a heat-transfer surface area of 1430 ft^2. The stale air blower is totally enclosed, for protection against dust and moisture. Cost may be in the neighborhood of $1000.

MUNTERS AB EXCHANGER

Address: PO Box 7093, S-191 07 Sollentona, Sweden. Makes rotary exchangers.

NORLETT EXCHANGER

Address: N-1801, Askim, Norway.

PASSIVE SOLAR PRODUCTS LTD. EXCHANGER

Address: 2234 Hanselman Ave., Saskatoon, Saskatchewan, Canada S7L 6A4. Tel.: (306) 244-7511. Some key persons: Lefebvre, Ron, Production Manager.

Earlier product: In 1980 the company was selling a "Humid-Fire Model 11" exchanger made elsewhere. This activity has been discontinued.

Early in 1982 the company was completing preparations for producing and selling two types of exchangers, called "Heat-Xchanger": a Model PA 150 and a Model PA 250.

Model PA 150

Overall dimensions: 48 in. x 36 in. x 11. Heat-exchange sheets are of plastic. Rate of airflow in each stream: 135 cfm. Efficiency: 72%. A drain for condensate is provided. The defrost system includes (at warm end of incoming air passage) a thermostat which, on sensing abnormally low temperature, turns off the fresh-air blower until the temperature returns to normal. The exchanger is fully prewired: just plug it in! Weight: 75 lb. Retail price will be about $695 Canadian, F.O.B. Saskatoon.

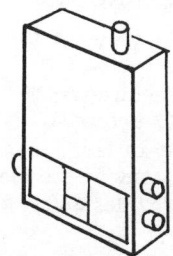

Extra-cost options include: Variable speed control, $25. Lint trap to permit use with washing machine.

Model PA 250

This exchanger, which is 44 in. x 36 in. x 14 in., provides 250 cfm with an efficiency of 71%. Weight: 100 lb. It is to be in production early in 1982. Price: about $980 Canadian.

The PA 150 is intended for use in average-size houses. The PA 250 is suitable for use in large houses or in indoor swimming pool areas, etc.

Note: This company is, in fact, a division of Softec Industries Ltd., which has the same address and same telephone number.

TEMP-X-CHANGER DIVISION EXCHANGER

(A division of United Air Specialists)

Address: 4440 Creek Rd., Cincinnati, OH 45242. Tel.: (513) 891-0400.

Representative in Massachusetts: Air Equipment Co., 56 Glenwood Ave., Hyde Park, MA 02136. Tel.: (617) 361-8490.

Makes a variety of air-to-air heat-exchangers mainly for use in commercial and industrial applications. The exchangers are called Temp-X-Changers. Most employ vertical, diamond-embossed, aluminum sheets spaced 0.44 in. apart in the larger exchangers and 0.38 in. apart in the smaller models. The airflow pattern is counterflow. The casings are of 22 ga. galvanized steel.

Flowrates from 500 to 8000 cfm are available.

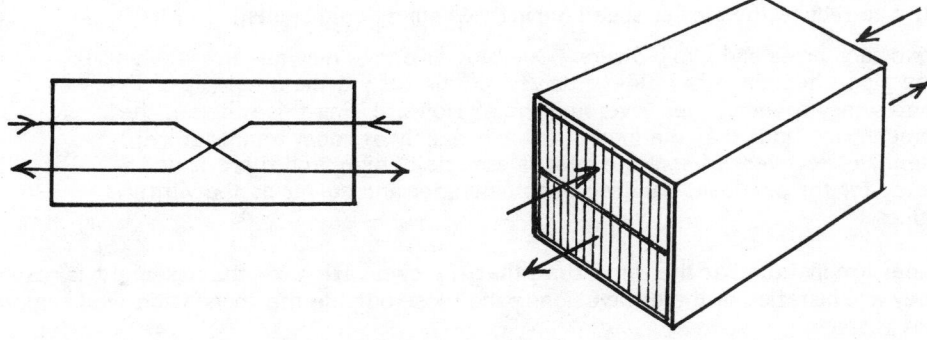

WING CO. (THE) EXCHANGERS

Address: The Wing Co. (Div. of Wing Industries, Inc.), 125 Moen Ave., Cranford, NJ 07016. Tel.: (201) 272-3600.

Dealer in Massachusetts: Emerson-Swan, Inc., 537 University Ave., PO Box 190, Norwood, MA 02062. Tel.: (617) 762-9000.

Makes a variety of large rotary air-to-air heat-exchangers for industry. Model names: Cormed, Correx, Enthalcorr. Many of the models include desiccant for recovery of latent heat. Smallest model (Correx Model 350) provides 660 cfm; and costs about $3500.

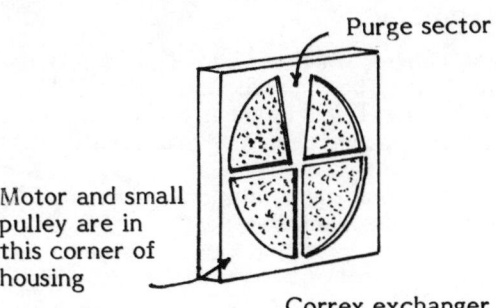

Purge sector

Motor and small pulley are in this corner of housing

Correx exchanger

DRAINAGE-PIPE-TYPE EXCHANGER

In this scheme, one form of which was suggested to me on 6/25/81 by Mark Kelley of Acorn Structures, Inc., use is made of a very common and inexpensive kind of drainage pipe, for example, a 4-in.-diameter corrugated polyethylene flexible pipe* costing about $70 per 250-ft. coil. The pipe is buried in the ground (at a depth of 5 or 6 ft., say) and thus remains at a temperature near 50°F. A small blower continually drives outdoor air into the house via this pipe, and accordingly the air that thus enters the house is at a temperature only a little below 50°F -- even if the outdoor temperature is 0°F.

Two possible improvements on the scheme are:

1. Turn the blower off when the outdoor windspeed is high. At such times, natural infiltration will provide all the fresh air needed. Turning off the blower, and thus stopping the flow of cold air in the pipe, reduces the tendency of the pipe and surrounding earth to cool down to, say, 40°F long before the winter is over. (Might pay also to reduce the blower speed during extremely cold spells.)

2. Use two such pipes and two blowers. One blower drives new air into the house via one pipe while the other blower drives old air out via the other pipe. Every 3 minutes they reverse, thus reversing the airflows. Using this scheme, the house occupant finds that the incoming air is nearly at room temperature; efficiency of recovery of sensible heat is especially high, and there is no tendency for the pertinent earth to become colder and colder as the winter continues.

If the pipes are installed at the same time that the excavation for the basement is made (and especially if they are installed in the excavation space close outside the foundation-wall region) the added labor cost is small.

Because some condensation may occur in the pipes, especially in summer, the pipes should be pitched slightly and a drain should be provided at the low end. Organic material may grow in the pipes and could, presumably, cause health problems.

*Such pipes are sold by A. D. S., Inc., by Hancor Inc., and others, I am told.

SAUNDERS EXCHANGER

<u>Model S Exchanger</u>

This exchanger, invented by Normal B. Saunders of 15 Ellis Rd., Weston, MA 02193 (Tel.: (617) 894-4748) employs no fans or blowers -- no electrical power at all. The two countercurrent airstreams are driven by the wind, with some assistance from chimney effect. Whenever there is at least a gentle wind, the exchanger performs satisfactorily. Where there is no wind, it practically ceases to function.

In January 1981 the exchanger was in developmental stage.

The exchanger proper is situated within a long slender box-like housing that is 16 ft. long, 10 in. high and about 22 in. wide. (The actual width may be less than 22 in., but space for installation should be at least 22 in. wide.) The box is mounted, typically, between the rafters of a sloping roof; thus it takes up no space that is already in use. The box is not insulated; the assumption is that the existing insulation on the underside of the roof will suffice.

The exchanger includes many plane parallel aluminum sheets (about 3/4 in. apart) which define many parallel airspaces. Fresh air travels in airspaces 1, 3, etc., and stale air travels (in opposite direction) in airspaces 2, 4, etc. The total area of heat-transfer surface is about 600 ft^2. At each end of the long slender box there are two 8-in.-dia. ports. Air is supplied to the exchanger by a special scoop. This may be mounted in such manner that it automatically turns so as to face into the wind. Stale air emerging from the exchanger travels upward through an exhaust turbine; wind turns the turbine and this helps the exhaust process. Chimney effect helps also.

When the windspeed is 15 mph, the rate of flow in each stream is about 350 cfm (predicted), the pressure drop is about 0.07 in. water (predicted), and the efficiency of sensible-heat recovery is about 50 to 70% (predicted).

It is expected that the price may be below $400.

The wind-drive system is covered by U.S. Patent 4, 296, 733.

<u>Model L Exchanger</u>

This is similar but employs heat-transfer sheets that are of plastic and are premeable to water vapor. Thus latent heat, as well as sensible heat, can be recovered.

<u>Note:</u> I learned sometime earlier, in about May of 1980, of a wind-powered air-to-air heat-exchanger proposed by John C. Wiles Jr. of Suitland, MD. The device is situated between floors and makes use of a very-large between-floors area. Both of the airstreams are driven by the wind; both flow generally eastward, say. The incoming air enters via a series of large-diameter pipes which are surrounded by the warm outgoing air. Thin (0.006-in.) horizontal sheets of polyethylene spread below the incoming-air pipes catch any condensate dripping from them. The scheme has obvious good features and obvious limitations.

Chapter 29

MEASUREMENT OF REPLACANCE

Introduction

Tracer gas method

Use of sulfur hexafluoride

Use of other tracer gases

Pressurization method of measuring leakiness

INTRODUCTION

Here I describe methods of measuring replacance, that is, the extent to which, in a given period of time, the old air in a house has been expelled and fresh air has been taken in. As explained in Chapters 7 and 8, this is a very different subject from the rate of fresh air input into the house.

    The methods of measuring replacance are applicable, of course, irrespective of whether the replacing of old air by new air is brought about by natural infiltration (helped by wind and chimney effect) or by simple venting fans or by an air-to-air heat-exchanger.

TRACER GAS METHOD

The tracer-gas method of measuring replacance starts with the sudden release, into the air in the house, of a small quantity of a special gas: a gas that is (a) different from all other gases present in room air, (b) easily detected even at very low concentrations, (c) stable, and (d) non-toxic. After releasing the tracer gas, the investigator uses one or more fans to distribute the gas uniformly throughout the house, i.e., to mix it thoroughly with all of the indoor air.

    He then measures the concentration of the tracer gas at the end of each 15-minute or 30-minute time interval. Each measurement is made on an air sample that is a composite of small amounts collected in different rooms and thus is representative of the indoor air as a whole. Successive measurements show, of course, successively lower concentrations of the tracer gas.

    The investigator then plots a graph of tracer-gas concentration as a function of time. From the graph he finds the replacance at the end of any desired period of time, such as one, two, or three hours. The one hour value, defined as the replacivity (see Chapter 7), is especially useful. Example: If, at the end of two hours, the tracer-gas concentration has decreased by 90%, the investigator knows that the two-hour replacance is 90%; 90% of the air that was in the house at the start of the two-hour period has left the house and has been replaced by fresh air.

    Finding the rate of fresh-air input is simple enough -- once the replacance has been evaluated -- if it can be assumed that all of the incoming air is promptly and thoroughly mixed with the air already in the house. All one has to do is consult the table shown on in Chapter 7. Alternatively, one may evaluate:

$$\frac{\text{(one house volume per hour)(41.6 minutes)}}{\text{half-life}}$$

where half-life means the time interval (in minutes) in which the tracer-gas concentration decreases by 50%. Example: An investigator finds that the tracer-gas concentration decreases to half in 60 minutes. What is the rate of fresh-air input? Answer: (one house volume per hour)(41.6/60) = 0.69 house volume per hour.

164

Various tracer gases may be used. Sulfur hexafluoride is, today, the usual choice. Therefore I discuss it first.

USE OF SULFUR HEXAFLUORIDE

Sulfur hexafluoride, SF_6, when dispersed in room air, can be detected even when present in very small amounts, such as a few parts per billion. Under normal conditions of use as tracer, this gas is stable, odorless, and non-toxic. Ordinarily there is none of it in room air. An extremely sensitive detection method is used: a method involving capture, by the SF_6 molecules, of electrons that have a certain energy. The equipment is called a gas chromatograph. A carrier gas -- argon -- facilitates the measurement process.

Before the test is begun, a preliminary test is made to see whether any freon is present in the room air. If any freon is present, it can greatly interfere with the detection of SF_6 inasmuch as these two materials have somewhat similar electron capture capabilities.

The main measurement is started by releasing about 10 cubic centimeters of SF_6 gas into the room air -- say a ground-floor room on the upwind side of the house. Fans are then used to distribute the tracer gas uniformly throughout the indoor air. The SF_6 concentration is then measured every 15 or 30 minutes, and, from a graph of the results, the investigator computes the replacance pertinent to any given time interval.

USE OF OTHER TRACER GASES

In most of the measurements made before the mid-1960s, helium (He) was used as tracer gas. It is detected, when mixed with room air, by measuring the increase in thermal conductivity of the air. Pure helium has a thermal conductivity about six times that of air, and even if room air includes only about 0.1% or 0.5% of helium, the slight increase in thermal conductivity can be detected.

Of the order of 0.1 to 1 lb. of helium is required for the test. This is unfortunate inasmuch as helium is in short supply.

Some investigators have used methane (CH_4), or radioactive krypton (^{85}Kr), or radioactive argon (^{41}A) or nitrous oxide (N_2O).

A survey of the various gases used has been published by the Lawrence Berkeley Laboratory of the University of California. See report LBL-8394 "An Intercomparison of Tracer Gases Used for Air Infiltration Measurements", by D. T. Grimsrud et al. (Bibl. G-810).

PRESSURIZATION METHOD OF MEASURING LEAKINESS

It is often important to obtain a rough idea of the replacivity of the air in a house that has no air-to-air heat-exchanger. A typical house allows much air to leak in and leak out, thanks to the existence of countless small openings (cracks, etc.) and perhaps a few larger openings. Therefore wind and chimney effect may cause considerable infiltration.

To evaluate the leakiness, or typical rate of natural infiltration, several groups have developed methods involving pressurization. An investigator forces air into the house by means of a blower. He adjusts the blower speed so that the rate of fresh air input is just enough to cause the indoor air pressure to exceed the outdoor air pressure by 50 pascal (which correspond to 0.2 in. of water). At the same time, he measures the rate of the blower-forced air inflow. This rate is usually assumed to be about equal to the natural rate of fresh-air inflow (for the house in question) when there is a 25-to-30-mph wind, I understand.

Typically, the blower is attached snugly to a lightweight plate (e.g., sheet of plywood) that is installed temporarily in place of an outside door. The plate is carefully sealed to the door frame.

Infiltrometer Made By Ener-Corp Management Ltd.

An "infiltrometer" based on the above-described principle and taking advantage of development work done by the Canadian National Research Council is produced and marketed by Ener-Corp Management Ltd. of 2 Donald St., Winnipeg, Manitoba, Canada R3L OK5. The infiltrometer includes a blower and also sensing and recording equipment that provides a detailed account of the infiltration characteristics of the building in question. (The company makes, also, an air-to-air heat-exchanger.)

Chapter 30

OTHER MEASUREMENTS

Introduction

Groups that have measured heat-recovery efficiency

Test method used by investigators at Princeton University

INTRODUCTION

I shall not attempt to list or describe the various methods of measuring pressure head, airspeed, volume flowrate, etc. Many kinds of instruments are available.

Included below are some remarks on groups that have developed equipment for measuring efficiency of sensible-heat exchange.

GROUPS THAT HAVE MEASURED HEAT-RECOVERY EFFICIENCY

I understand that heat-recovery-efficiency tests (on air-to-air heat-exchangers) have been made at:

University of California at Berkeley, Lawrence Berkeley Laboratory, Building Ventilation and Indoor Air Quality Program, William Fisk, Project Leader of Heat Exchanger Project

Prairie Agricultural Machinery Institute, Humbolt, Sask., Canada.

Princeton University, Center for Energy and Environmental Studies.

I am told that different groups use different test conditions and test methods -- and may arrive at somewhat different results. Apparently there is a need to determine what sets of conditions are most pertinent and which test methods are most accurate. Meanwhile the published values of efficiency should be regarded as being tentative and subject to some revision.

TEST METHOD USED BY INVESTIGATORS AT PRINCETON UNIVERSITY

A. Persily et al at the Center for Energy and Environmental Studies at Princeton University (Engineering Quadrangle, Princeton University, Princeton, NJ 08544) have been testing the sensible-heat-recovery efficiency of an air-to-air heat-exchanger with the aid of a small, very well insulated, very airtight, test room that simulates a small house. Electric heaters are used to hold the room at a fixed temperature head relative to the temperature of the surrounding air. Test runs are made with (1) no air change, (2) air change provided by a simple blower-and-vent system, and (3) air change provided by an air-to-air heat-exchanger. Changes in efficiency are reflected in changes in amount of electrical power needed to maintain a fixed temperature head.

Preliminary tests on one well-known, small exchanger that employs crossflow revealed, at first, some leakage of air from one airstream to the other. After this had been corrected, efficiencies close to, or perhaps slightly lower than, the manufacturer's published values were obtained.

Chapter 31

ECONOMICS OF EXCHANGER USE

Introduction

Simple calculation of saving during first year

General formula for saving during first year

Saving during 5-year period assuming 15% annual increase in cost of auxiliary heat

Ten-year saving

Some possible bonuses

INTRODUCTION

Here a very simple subject is discussed: how much money does a homeowner save by employing an air-to-air heat-exchanger to freshen the air in his house -- compared to merely using forced ventilation with no recovery of heat from the outgoing air?

The answer is: a great deal, if the house is in a cold climate and a high rate of fresh-air input (such as 200 cfm) is desired. Under such circumstances an air-to-air heat-exchanger "pays for itself" in about 3 to 7 years.

SAMPLE CALCULATION OF SAVING DURING FIRST YEAR

It is self-evident that if, throughout the first year of use of a certain heat-exchanger (which has a sensible-heat-recovery efficiency of 75%), this exchanger supplies 200 ft^3 of fresh air each minute to a house in a 5000 F degree-day location, the amount of heat recovered is:

$$(200 \text{ ft}^3/\text{min.})(60 \text{ min./hr})(24 \text{ hr/day})(0.074 \text{ lb/ft}^3)(0.24 \text{ Btu/lb } ^\circ\text{F})(5000 \text{ F DD})(75\%)$$

= 19.2 million Btu per winter.

If the same amount of fresh air had been supplied with no heat recovery, and if a 70% efficient oil furnace (using oil at \$1.20/gal.) had been used to heat this air to 65 $^\circ$F (i.e., with a cost of \$12 per million Btu), the expenditure would have been 19.2 x \$12 = \$230.

Thus, in an important sense, about \$230 would have been saved, during this first winter, by the exchanger. (Relative to using no exchanger and providing no fresh air, there is no monetary saving at all. However, the amount of discomfort to the house occupants, and the damage to their health, might then be enormous.)

Warning Concerning Choice Of Degree-Day Figure

Air-to-air heat exchangers are especially useful in tightly built houses. Usually, such houses are well insulated and may require little or no auxiliary heat as long as the outdoor temperature exceeds 55°F say (or even lower temperature for houses that are especially well insulated and receive much heat from light bulbs, cooking stove, etc. and from solar radiation). In other words, the heat saved by an air-to-air heat-exchanger when the outdoor temperature exceeds 55°F, say, is not needed at that time, in such a house, and so has little or no monetary value. In estimating the dollar saving by the exchanger, one should use a degree-day values based -- not on 65°F -- but on 55°F or whatever temperature is appropriate.

Warning Concerning Overall Cost Of Exchanger

A person considering buying an exchanger should consider not only the F.O.B. purchase cost but also (1) the cost of shipping and installing the exchanger, (2) the annual operating cost (cost of electrical power) usually about 1/10 or 1/5 of the saving and thus nearly negligible, and (3) the annual maintenance cost, which is negligible ordinarily.

Saving Relative To Use Of Gas Or Electricity

Relative to the use of a 70% efficient gas furnace, with gas priced at $4 per 1000 ft^3 or $5.7 per million Btu, the saving (in the above-stated example) would be about $113, i.e., about half as great. Relative to the use of electricity at 7¢/kWh, or $20.6 per million Btu, the saving would be about twice as great as with oil, i.e., about $400.

GENERAL FORMULA FOR SAVING DURING FIRST YEAR

The general formula is: Dollar saving during the first winter (from use of the given exchanger with given sensible-heat-recovery efficiency) is the product ABCDEF where:

A = (rate of fresh air input, in ft^3/min)(60 min/hr)(24 hr/day)

B = 0.074 lb/ft^3

C = 0.24 Btu/(lb $^\circ$F)

D = degree day value

E = efficiency of sensible-heat recovery

F = cost of auxiliary heat in dollars per Btu.

Combining all of the above-listed constants into a single constant, one finds that the formula boils down to this:

$$\text{Dollar saving in first year} = \left(25.6 \; \frac{\text{Btu min}}{^\circ\text{F day ft}^3}\right) \left(\begin{array}{c}\text{rate of air}\\ \text{input, ft}^3\text{/min}\end{array}\right) \left(\text{DD}\right) \left(\text{Efficiency}\right) \left(\begin{array}{c}\text{Dollars}\\ \text{per Btu}\end{array}\right)$$

Or, very briefly: $\boxed{\text{Saving} = (25.6)(\text{cfm})(\text{DD})(\text{Eff.})(\$/\text{Btu})}$

SAVING DURING 5-YEAR PERIOD ASSUMING 15% ANNUAL INCREASE IN COST OF AUXILIARY HEAT

If the cost of auxiliary heat escalates at 15% per year, then the saving during the first five years of use of an exchanger is about 35% more than would be the case if there were no escalation. Using, as first-year costs of auxiliary heat, the figures given above (namely $12, $5.7, and $20.6 per million Btu using oil, gas, and electricity), the five year saving with 15% escalation, for the specific case discussed on the previous page, is $1550, $750, and $2670 respectively.

These are big numbers! Several times the cost of the exchanger!

TEN-YEAR SAVING

Here, again assuming 15% escalation in cost of auxiliary heat, one finds the saving to be about $4670, $2220, and $8030 respectively.

These are very big numbers!

The saving is about twice as great as would be the case if there were no escalation.

SOME POSSIBLE BONUSES

If the exchanger provides transfer of heat only, the house occupant may enjoy these bonuses:

In winter, the load on the furnace is reduced and accordingly a smaller cheaper furnace may suffice. Or perhaps the furnace may be dispensed with entirely; perhaps a small electrical heater can provide all the auxiliary heat needed.

In summer, the load on the air conditioning system is reduced. A smaller cheaper conditioner may suffice. Possibly the occupants can get by with no conditioner.

If the exchanger transfers both heat and water, humidification needs in winter are reduced and dehumidification needs in summer are reduced.

Federal and state governments may provide certain tax benefits to persons who install heat-exchangers.

Added Note

An excellent treatment of the cost-effectiveness of air-to-air heat-exchangers may be found in Energy Sourcebook (Bibl. M-97), pages 592 - 597, and in a report by G.D. Roseme et al of Lawrence Berkeley Laboratory, Univ. of Calif. (Bibl. U-520-13).

Chapter 32

BUYING AN EXCHANGER

Before buying an air-to-air heat-exchanger, the householder should address these questions:

What kinds of pollutants are present in my house? Are they sufficiently concentrated to be health hazards or to produce discomfort? What are their sources? Can they be eliminated?

Do I really need an exchanger? Or is my house leaky enough so that, nearly always, natural infiltration provides all the fresh air needed? If not, can I get by merely by opening a window or two?

If I buy an exchanger, am I doing so primarily to supply fresh air, or primarily to avoid excessive humidity? Is the recovery of latent heat (from moist indoor air) a major goal? Do I intend to use the exchanger in summer also?

Is a fresh air supply needed in just one room? Or in several rooms? If the latter, will I feel obliged to install a duct system that will serve several rooms? Would it be better to buy several small exchangers and put one in each room -- and avoid the need for ducts?

What capacity is needed? Will 60 or 100 cfm suffice? Or is a higher capacity needed?

Is the climate cold enough so that frost formation and clogging of passages in the exchanger could be serious threats?

How elaborate a control system is wanted? Is a versatile and automatic system necessary? Or will manual controls suffice?

Where, exactly, would the exchanger be installed? Will there be adequate access for installing it and, subsequently, inspecting and maintaining it?

Who will install it and who will take the overall responsibility for seeing that it performs properly?

What price is acceptable? What annual operating cost is acceptable?

Will the exchanger be truly cost-effective?

Having considered these questions, the householder should decide what type of exchanger he wants: rotary, or fixed with laminar flow, or fixed with turbulent flow. Also he should decide which specific make and model is most appropriate, whether a suitable dealer is near enough at hand, whether price, delivery period, and warranty are satisfactory.

Of course, the householder would do well to consult with available friends and experts who have had experience with exchangers.

The following table summarizes the specifications, prices, etc., of most types of air-to-air heat-exchangers for houses.

SPECIFICATIONS SUMMARY showing product name, overall dimensions, weight, flowrate (and alternative flowrates), efficiency, blower power, and price (all subject to change)

Berner International
Econofresher GV-120 21½x12x8¾ in. 21 lb. 30(60)cfm 82%(75%)enth. 25(40)w $350. Rotary type.
Economini 500 53x27x14 in. 134 lb. 160(225,295)cfm 78(73,70)% 160(220,270)w $1200. Rotary.

Q-Dot
TRU 120M-6A 22.8x9x9.3 in. 100 lb. - - - $400. Heat-pipe type. Blowers etc. not included.

Mitsubishi Electric Sales America
Lossnay VL-1500 22x15x12 in. 29 lb. 28(50,71)cfm 76(73,70)% 22(40,56)w $310. Wall-mounted.
Lossnay VL-1500-Z 22x15x12 in. 29 lb. 40(62)cfm 64(60)% 40(56)w $350. Ceiling-mounted.
Lossnay VL-500-B2 13x10x9 in. 6.2 lb. 23 cfm 65% 30w $110. Mounted on bathroom wall.
Lossnay VL-1500-V Much like VL-1500 but intended to be mounted with long axis vertical.

Des Champs Laboratories
Z-Duct 79M.4-RU 62x16x14 in. 75 lb. 150-200 cfm 85% at 150 cfm 190 w at 150 cfm. $450.
Z-Duct 79M.2-RU 58x16x10 in. 55 lb. 75 cfm 85% 100w $400. Ready early in 1982.
Z-Duct 79M.6-RU 69x16x20 in. 100 lb. 350 cfm 85% 380w $600. Ready early in 1982.

Enercon
Model 150 68 in. long, 20 in. dia. 70 lb. 70(200)cfm 65(65)% 105w $1500.
Model 200 80 in. long, 16 in. dia. 70 lb. 70(200)cfm 65(65)% 105w - Agricultural use also.
Model 400 80 in. long, 20 in. dia. 80 lb. 220(350)cfm 65(65)% 500w - Agricultural use also.

Conservation Energy Systems
VanEE R200 48x15x22 in. 70 lb. 0 - 200 cfm 71% at 200 cfm 200w $700. Continuously variable flowrates.

Air Changer Co.
Deluxe 54x22x14 in. 70 lb. 0 - 150 cfm 92(89,85)% at 50(100,150)cfm 70w at 150 cfm. $760.
Standard 54x22x14 in. 70 lb. 100 cfm 89% 70w $670. Has few controls.
Basic 54x22x14 in. 60 lb. - - $400. Blowers and controls not included.

Passive Solar Products
HeatXchanger Model PA 150 44x36x11 in. 75 lb. 135 cfm 72% 125w $600.
HeatXchanger Model PA 250 48x36x14 in. 100 lb. 250 cfm 71% 250w $800. Ready early in 1982.

Temovex AB
Model 480-S 65x14½x17½ in. 120 lb. 75(150)cfm 82(79)% 95(190)w. About $1000.
Cabinet type. 83x24x17 in. 220 lb. Same cfm, eff., power as above. $1100. Mounted vertically in laundry room.

Aldes Riehs U.S.A.
Aldes VMPI 57x18x11 in. 50 lb. 60(230)cfm 73(68)% 100(220)w About $750. Blowers not included.

Bahco Ventilation
Minimaster 40x24x11 in. 90 lb. 100 cfm 60 to 65% 170w Mounted above cooking stove.

Advanced Idea Mechanics AIM-120 22x16x16 in. 55 lb. 120cfm 76% 150w $550. Ready early in 1982.
Ener-Corp Management Std. Mod. 70x16x14 in. 80 lb. 75(95,130)cfm 95(86,80)% 340w at 130cfm. $900. Early 1982.
The Memphremagog Group Residential Mod. 50x28x14 in. 85 lb. 100-150cfm 80% 65w $500 to $800.

Appendix 1

UNITS AND CONVERSIONS

Energy   1 J   = exactly 1 newton-meter
i.e., 10^5 dyne-meter,
i.e., 10^7 erg
i.e., 1 watt-second
= 2.778×10^{-7} kWh
= 9.478×10^{-4} Btu

1 kWh   = exactly (3600)(1000) watt-second
i.e., 3.6×10^6 J
i.e., 3.6×10^{13} erg
= 3.412×10^3 Btu

1 Btu   = exactly 1 (lb. H_2O at $63^\circ F$)($^\circ F$)
= 1.055×10^3 J
= 1.055×10^{10} erg
= 2.931×10^{-4} kWh
= 778.2 ft lb.

1 therm   = exactly 10^5 Btu

1 42-gal. barrel of oil = 5.8×10^6 Btu

1 Cubic foot of natural gas = 1031 Btu

1 lb. coal = 1.25×10^4 Btu

Length   1 ft. = 0.305m

Area   1 ft^2 = 0.0929 m^2

Volume   1 ft^3 = 0.0283 m^3

Mass   1 lb = 0.4536 kg

Force   1 pound force = 4.448 N

Pressure   1 pound force/in^2 = 6.9 kN/m^2
= 6.9 kPa

1 atm. = 14.7 pound force/in^2 = 101.3 kPa = 407 in. H_2O

Viscosity   (i.e., absolute viscosity, dynamic viscosity)

1 poise = 1 P = 1 gram/(sec. cm.) = 100 centipoise

1 pound force, sec/ft^2 = 0.047 Pa sec.

Viscosity in centipoise x 0.000672 = viscosity in lb/(sec. ft)

Viscosity in centipoise x 0.0000209 = viscosity in (lb force)(sec)/ft^2

Example for air at $70^\circ F$ and atm. pressure: viscosity is 0.018 centipoise, 0.000,000,375 (lb.force x sec)/(ft^2), 0.000,0121 lb. mass/(sec. ft.)

Heat flowrate   1 Btu/hr = 0.293 W

Thermal conductivity   1 Btu/(ft hr $^\circ F$) = 1.730 W/(m$^\circ$C)

Coefficient of heat transfer: 1 Btu/(ft^2 hr $^\circ F$) = 5.678 W/(m^2 $^\circ$C)

Specific heat capacity   1 Btu/(lb. $^\circ F$) = 4.186×10^3 J/(kg $^\circ$C)

Enthalpy   1 Btu/lb. = 2.326 J/g

Sources: Various, including Reay (R-25) p. 574, 575.

Appendix 2

BIBLIOGRAPHY

A-52   Air Infiltration Centre, "Air Infiltration Review", a quarterly published from Old Bracknell Lane, Bracknell, Berkshire, RG 12 4AH, England.

B-251a Besant, R. W., R. S. Dumont, and D. Van Ee, "An Air-to-Air Heat-Exchanger for Residences", an engineering bulletin published in 1978 by the Extension Division, University of Saskatchewan, Saskatoon, Sask., Canada S7N OWO. $2.

B-330   Blomsterberg, A. K., and D. T. Harrje, "Approaches to Evaluation of Air Infiltration Energy Losses in Buildings", ASHRAE Transactions 85. (1979), Part 1. 18 p.

C-205   Chapman, A.J., "Work Heat Transfer", Macmillan Publishing Co., Inc., 3rd ed.

C-595   Committee on Industrial Ventilation, "Industrial Ventilation: a Manual of Recommended Practice", 16th ed., (1980). $10. Published from PO Box 16153, Lansing, MI 48901.

C-720   Coren, Gerald, "Chemical Engineering", ARCO Publishing Co. (1980).

D-400   Dockery, D. W., and J. D. Spengler, "Indoor-Outdoor Relationships of Respirable Sulfates and Particles", Atmosph. Environment 15, 335 (1980). Derives formulas for concentration as a function of rate of air change.

F-70   Fisk, W. J., G. D. Roseme, and C. D. Hollowell, "Test Results and Methods: Residential Air-to-Air Heat Exchangers for Maintaining Indoor Air Quality and Saving Energy", LBL Report 12280. University of California at Berkeley, (1981).

F-700   Fuller, Winslow, "What's in the Air for Tightly Built Houses?", Solar Age, June 1981, p. 30.

G-810   Grimsrud, D.T., et al: "An Intercomparison of Tracer Gases Used for Air Infiltration Measurements", LBL Report 8394. University of California at Berkeley. (1979).

H-156   Hollowell, C. D., et al: "Building Ventilation and Indoor Air Quality Program", LBL Report 9284. (1979). Chapter from an annual report of 1978. University of California at Berkeley.

H-157   Hollowell, C.D., et al: "Building Ventilation and Indoor Air Quality". LBL Report 10391. (1980). University of California at Berkeley.

I-430   International Symposium on Indoor Air Pollution, Health and Energy Conservation, Oct. 13-16, 1981, at Amherst, Mass. "Extended Summary of Papers". Symposium organized by two schools at Harvard University, Cambridge, MA 02138: School of Public Health, Kennedy School of Government. Key organizer: Prof. John D. Spengler of the Harvard School of Public Health. Report issued on Oct. 13, 1981, to symposium participants. Out of print thereafter. About 600 p.

J-110   Jakob, Max, "Heat Transfer", Wiley & Sons, New York, NY. (1957).

K-52   Kays, W. M., and A. L. London, "Compact Heat Exchangers", McGraw-Hill Book Co., New York, NY. (1964). 2nd ed.

K-54   Kays, W.M. and M.E. Crawford, "Convective Heat and Mass Transfer", McGraw-Hill Book Co., New York, NY. (1980). 420 p.

K-400   Knudsen, J. G., and D. L. Katz, "Fliud Dynamics and Heat Transfer", McGraw-Hill Book Co., New York, NY (1958).

M-86   McAdams, W. H., "Heat Transmission", McGraw-Hill Book Co., New York, NY. (1942). 2nd ed.

M-97   McRae, Alexander (et al), Editors, "Energy Sourcebook", published by Aspen Systems Corp., Germantown, MD. (1977). See pages 551 - 606.

N-24m   National Academy of Sciences, National Academy Press, "Formaldehyde", (1981). 340 p. Paperback. ISBN 0-309-03146-X. $11.50.

N-25   National Academy of Sciences, National Academy Press, "Indoor Pollutants", Oct. 1981. 560 p. Paperback. By Committee on Indoor Pollutants, $16.25. Available from 2101 Constitution Ave., NW, Washington, D.C. 20418. ISBN 0-309-03188-5.

N-200   Nazaroff, W. W., et al: "The Use of Mechanical Ventilation with Heat Recovery for Controlling Radon and Radon-Daughter Concentrations", LBL Report 10222. (1980). University of California at Berkeley.

0-145   Offermann, F. J., J. R. Girman, and C. D. Hollowell, "Midway House-Tightening Project: A Study of Indoor Air Quality", Lawrence Berkeley Laboratory, Report LBL 12777. May 1981. 30 p. University of California at Berkeley.

0-150   O'Gallaghan, P. W., "Building for Energy Conservation", Pergamon Press, (1978). 231 p. ISBN 0-08-022120-3.

R-25   Reay, D. A., "Heat Recovery Systems: A Directory of Equipment and Techniques", E. & F. N. Spon, London. (1979). 600 p. ISBN 0-419-11400-9.

S-45   Schubert, R. C., and L. D. Ryan, "Fundamentals of Solar Heating", Prentice-Hall, Inc., Englewood, NJ 07632. (1981). 335 p. Hardbound. $25.95.

S-55   Scott, Linda A., "Indoor Air Pollution in Passive Structures", (1980). 10 p. St. Olaf College, Northfield, MN 55057.

S-146   Shapiro, Jacob, "Radiation Protection: A Guide for Scientists and Physicians", Harvard University Press, Cambridge, MA 01238. (1981). 500 p. Hardbound. $25.

S-162   Sheet Metal and Air Conditioning Contractors National Association, Inc., "Energy Recovery Equipment and Systems: Air-to-Air", (1978) About 100 p. Published from 8224 Old Courthouse Rd., Tysons Corner, Vienna, VA 22180. $22.50.

S-235gg Shurcliff, W. A., "Superinsulated Houses and Double-Envelope Houses", Brick House Publishing Co., 34 Essex St., Andover, MA 01810. (1981). 165 p. Paperback. $12.

S-235mm Shurcliff, W. A., "A New Concept in Air-to-Air Heat-Exchangers: Exchanger That Can Be Switched Instantly From Water-Recovering Mode to Water-Ejecting Mode: The DM-10 Hexar", 8 p. report of 12/12/81. $2. Available from 19 Appleton St., Cambridge, MA 02138.

S-363m   Spengler, J. D., D. W. Dockery, W.A. Turner, J.M. Wolfson, and B.C. Ferris, Jr., "Long-Term Measurements of Respirable Sulfates and Particles Inside and Outside Homes," Atmospheric Environment 15, 23 (1980).

U-471m   United States National Bureau of Standards, "Waste Heat Management Guidebook", NBS Handbook 121, 1977. U.S. Government Printing Office, Stock No. 003-003-01669-1. $2.75.

U-475-HE-40 U. S. Dept. of Energy, Solar Energy Research Institute, "Dehumidification in Passively
Cooled Buildings", by F. Arnold, R. Barlow, and K. Collier. May 1981. 50 p. Report SERI/
TR-631-995.

U-520-13 University of California, Lawrence Berkeley Laboratory, "Residential Ventilation with Heat
Recovery: Improving Indoor Air Quality and Saving Energy", G. D. Roseme et al, LBL
Report 9749. (1980). 29 p.

W-450 Woods, J. E., E. A. B. Maldonado, and G. L. Reynolds, "Safe and Energy Efficient Control
Strategies for Indoor Air Quality", published by Engineering Research Institute, Iowa State
University, Ames, IA 50011. ERI-BEUL 81-01. 38 p. Stresses need for better sensors and
controls. Includes big bibliography.

Z-75 Zamm, A. V., and Robert Gannon, "Your House Can Make You Sick!", Rotarian Magazine,
July 1981. p. 32.

Z-200 Zarling, J. P. "Air-to-Air Heat Recovery Devices for Residential and Small Commercial
Applications". 25-p. report of 1980 or 1981. From Dept. of Mech. Eng'g., University of
Alaska, Fairbanks, AK 99701.

Appendix 3

INDEX OF TRADE NAMES OF PRODUCTS

Trade Name	Pertinent Company
Air Changer Air-X-Changer	Air Changer Co. Ltd. (The); also Allen-Drerup-White, Ltd.
Cormed Correx	Wing Co. (The)
D. C.	D. C. Heat Exchangers, Ltd.; Conservation Energy Systems, Inc.
Del-Air	Del-Air Systems Ltd.; also D. C. Heat Exchangers, Ltd.
Econofresher	Berner International Corp.
Enthalcorr	Wing Co. (The)
GV-120	Berner International Corp.
HRU	Gaylor Industries, Inc.
Heat-Xchanger Humid-Fire No. 11	Passive Solar Products Ltd.
Lossnay	Mitsubishi Electric Corp.
MAC 1000	Ener-Corp Management Ltd.
Minimaster	Bahco Ventilation Ltd.
Model 11	Passive Solar Products Ltd.
Model 74 Model 79	Des Champs Laboratories, Inc.
PA 150	Passive Solar Products Ltd.
R-200	Conservation Energy Systems, Inc.
Senex Sharp	Berner International Corp.
Temovex 480A	Temovex AB
VanEe	D. C. Heat Exchangers, Ltd.; Conversation Energy Systems, Inc.
VMC	Genvex Energiteknik A/S
VMPI	Aldes Corp.; also Aldes Riehs U.S.A.
Z-Duct	Des Champs Laboratories, Inc.

Appendix 4

DIRECTORY OF ORGANIZATIONS

Advanced Idea Mechanics Ltd., 26 Ski Valley Crescent, London, Ont. N6K 3H3, Canada.

Air Changer Co. Ltd. (The), 334 King St. East, Studio 505, Toronto, Ont., Canada M5A IK8. Tel.: (416) 863-1792. White, Elizabeth. President. See also: Allen-Drerup-White Ltd.

Air Infiltration Centre, Old Bracknell Lane, Bracknell, Berkshire RF 12 4AH, Great Britain. Publishes Air Infiltration Review, a quarterly. Jackman, Peter J. Head. Liddament, Martin W. Senior scientist.

Aldes Riehs U.S.A., 157 Glenfield Rd., Sewickley, PA 15143. Tel.: (412) 741-2659. Imports Aldes VMPI air-to-air heat-exchanger from Aldes Co. in Lyon, France.

Allen-Drerup-White, Ltd., 334 King St. East, Studio 505, Toronto, Ont., Canada M5A IK8. Tel.: (416) 863-1792.
Allen, Greg. A principal. The company developed the exchanger called Air Changer to be produced and sold by The Air Changer Co. Ltd.

Automated Controls & Systems, 500 East Higgins Rd., Elkgrove, IL 60007. Tel.: (312) 860-6860. Makes exchanger cores. Kollock, Ray: a key person.

Bahco Ventilation Ltd., Bahco House, Beaumont Rd., Banbury, Oxfordshire, England OX16 7TB. Makes heat exchangers for cooking stoves etc.

Berner International Corp., 12 Sixth Rd., Woburn, MA 01801. Tel.: (617) 933-2180. Makes exchangers that employ rotors. Is US distributor for Japanese-made Sharp Corp. exchangers. Fischer, John: General Manager. Eriksson, Krister: Head of Engineering.

Blackhawk Industries, Inc., 607 Park St., Regina, Sask., Canada S4N 5N1. Tel.: (306) 924-1551. Supersedes Enercon Industries Ltd.

Cargocaire Engineering Corp., 79 Munroe St., Amesbury, MA 01913. Tel.: (617) 388-0600. Makes exchangers of fixed type and rotary type.

Conservation Energy Systems, Inc., Box 8280, Saskatoon, Sask., Canada S7K 6C6. Tel.: (306) 665-6030. Olmstead, Rick. President, Wholesale distributor of Van Ee exchangers made by D. C. Heat Exchangers, Ltd.

D. C. Heat Exchangers, Ltd., Box 8339, Saskatoon, Sask., Canada S7K 6C6. Tel.: (306) 384-0208. Van Ee, Dick. President. Makes Van Ee exchangers for houses and Del Air exchangers for animal barns; the exchangers for houses are sold at wholesale by Conservation Energy Systems, Inc., and the exchangers for animal barns are sold at wholesale by Del-Air Systems, Ltd.

Del-Air Systems, Ltd., Box 2500, Humboldt, Sask., Canada SOK 2AO. Tel.: (306) 682-5011. Key persons: Benning, Larry. Also Nevecosky, Edward, Secretary-Treasurer. Wholesale distributor of Del-Air exchangers (for animal barns) made by D. C. Heat Exchangers, Ltd.

Des Champs Laboratories Inc., PO Box 348, East Hanover, NJ 07936. Tel.: (201) 884-1460. Des Champs, Dr. Nicholas H.: President. Wyckoff, John T.: Production Manager for Residential Equipment. Makes many kinds of exchangers including Z-Duct exchangers.

Dravo Corp., PO Box 9305, Pittsburgh, PA 15225. Tel.: (412) 777-5932. Dormetta, Ron: Sales Manager. Makes industrial-type exchangers.

(Enercon Building Corp.: see following item.)

Enercon Consultants Ltd., 3813 Regina Ave., Regina, Sask., Canada S4S OH8. Schell, Michael B.: Marketing Research Head. The company is an affiliate of Enercon Building Corp. and Enercon Industries Ltd.

Enercon of America, Inc., 2020 Circle Dr., Worthington, MN 56187. Tel.: (507) 372-2442. This company is a USA subsidiary of the Canadian company Enercon Consultants Ltd.

Enercon Industries Ltd., 2073 Cornwall St., Regina, Sask., Canada S4P 2K6. Tel.: (306) 585-0025. Affiliated with several above-listed companies. Markets Enercon Fresh Air Heat Exchangers.

Ener-Corp. Management Ltd., 2 Donald St., Winnipeg, Manitoba, Canada R3L OK5. Tel.: (204) 447-1283. Makes infiltrometer and heat-exchanger. Giesbrecht, Peter. Key engineer.

Equinox, Inc., Granite Block, Peterborough, NH 03458. Fuller, Winslow. Writes on indoor pollution.

Gaylord Industries, Inc., PO Box 558, 9600 SW Seely Ave., Wilsonville, OR 97070. Tel.: (503) 682-3801. Makes exchangers for kitchen stove exhausts.

Genvex Energiteknik A/S, Blegdamsmej 104, DK-2100, København ∅, Denmark. Makes <u>VMC</u> exchanger.

Harvard University, Cambridge, MA 02138.
  Abernathy, Prof. Frederick H. Dept. of Applied Physics. Heat conservation.
  Shurcliff, Dr. William A. Physics Dept. (Home address: 19 Appleton St., Cambridge, MA 02138). Solar heating. Energy conservation.
  Spengler, Dr. John D. Harvard School of Public Health. Indoor pollution.
    Co-Chairman of the "Committee on Indoor Pollution" of the National Academy of Sciences.
  Rudnick, Dr. Stephen. Harvard School of Public Health. Radon pollution.

(Hastings Industries: see Dravo Corp.)

Heatex AB, Hammarsvagen, Sweden. Makes exchanger cores.

Hydrothermal Ltd., Adelaide, Australia. Makes cooling-type exchangers that employ water sprays.

Iowa State University, Engineering Research Institute, Ames, IA 50011. Department of Mechanical Engineering: Maldonado, E. A. B., Reynolds, G. L., and Woods, J. E. Experts on pollutants, exchangers, sensors, controls.

Long Equipment Co., PO Box 0108, Holbrook, MA 02343. Tel.: (617) 767-3595. Massachusetts representative of Dravo Corp.

(Melco Sales, Inc.: see Mitsubishi Electric Sales America, Inc.)

Memphremagog Group, PO Box 456, Newport, VT 05855. Tel.: (802) 334-8821. Hamilton, Blair, and Lepage, Armand: Key persons concerning exchanger design.

Mitsubishi Electric Corp., Japan.
  Lossnay Engineering Section, Nakatsugawa Works, Nakatsugawa, Gifu Pref., Japan. Yoshino, Masataka, Section Manager.
  Overseas Marketing, Group C. 2-3, Marunouchi 2-chome, Chiyoda-Ku, Tokyo, Japan. Kondo, Kyo: Assistant Manager of Group.
  American sales headquarters: see Mitsubishi Electric Sales America, Inc.
  Representative in New England: see R.F. Walker Associates.

Mitsubishi Electric Sales America, Inc. (MELCO), 3030 E. Victoria St., Compton, CA 90221. Tel.: (213) 537-7132. Also (800) 421-1132.
  Thomas, Mike: Manager of Sales of Mitsubishi Exchangers.

Munters, A. B., PO Box 7093, S-191 07 Sollentona, Sweden. Makes exchangers of rotary type.

National Research Council of Canada, Division of Building Research, Prairie Regional Station, Saskatoon, Sask., Canada S7N OW9. Dumont, Robert S. Heat exchanger design.

(Northeast Equipment Co.: see R. F. Walker Associates)

Passive Solar Products Ltd., 2234 Hanselman Ave., Saskatoon, Sask., Canada S7L 6A4. Tel.: (306) 244-7511. Makes HeatXchanger: PA 150, PA 250, etc.

Princeton University, Center for Energy and Environmental Studies, Engineering Quadrangle, Princeton, NJ 08544. Persily, Andy, and Dutt, G. S.: studies of air change and of exchanger performance.

Q-Dot Corp., 151 (or 726) Regal Row, Dallas, TX 75247. Tel.: (214) 630-1224. Bucher, Axel. A key person. Company makes small and large exchangers that employ heat-pipes.

(R. F. Walker Associates: see "Walker, R. F., Assoc.")

St. Olaf College, Northfield, MN 55057. Scott, Dr. Linda A. Tel.: (507) 663-3102. Indoor pollution.

Saskatchewan Department of Mineral Resources, 1404 Toronto-Dominion Bank Bldg., Regina, Sask., Canada S4P 3P5. Eyre, David, Energy conservation and exchanger design.

Sharp Corp., Japan. (See also Berner International Corp., which is the US distributor of Sharp Corp. rotary-type exchangers for houses.)

Sheet Metal and Air Conditioning Contractors National Assn., Inc., 8224 Old Courthouse Rd., Tysons Corner, Vienna, VA 22180. Tel.: (903) 790-9890. Published 1978 book on air-to-air heat-exchangers. See Bibl. S-162.

(Softech Industries, Ltd.: see Passive Solar Products Ltd.)

Stanford University, Stanford, CA.
London, A. Lewis, and Kays, W.M.: experts on heat-exchangers, and authors of book on subject.

Svenska Flakt Fabriken, Sweden.

Temovex AB, Box 111, S 265 01 Astorp, Sweden. The "visiting address" is Bangatan 5, Astorp. Tel.: 042 550 20. Templer, Borje: Manager.

Temp-X-Changer Division (of United Air Specialists), 4440 Creek Rd., Cincinnati, OH 45242. Tel.: (513) 891-0400.

(The Air Changer Co. Ltd.: see Air Changer Co. Ltd. (The))

(United Air Specialists: see Temp-X-Changer Division)

United States Government. Solar Energy Research Institute, 1536 Cole Blvd., Golden, CO 80401. Holtz, Michael, has explored heat-exchanger technology.

University of Alaska, Fairbanks, AK 99701.
Zarling, Prof. John P. Dept. of Mech. Eng'g. Exchanger technology.

University of California, Lawrence Berkeley Laboratory, Berkeley, CA 94720.
Fisk, W. J., Grimsrud, C. D., Roseme, G. D., and Rosenfeld, A. H.: energy conservation, heat-exchangers, surveys, tests, analyses.

University of Saskatchewan, Saskatoon, Sask., Canada S7N OWO. Dept. of Mechanical Engineering. Besant, Robert W., Dumont, R. S., Van Ee, Dick. Energy conservation, heat-exchangers. Special address of R. S. Dumont: Div. of Bldg. Research, 110 Gymnasium Rd., Saskatoon, Sask., S7N OW9.

University of Wisconsin at Madison, College of Engineering, 1500 Johnson Dr., Madison, WI 53706. Mitchell, Prof. J. W. Heat-exchangers.

Wing Co. (The), 125 Moen Ave., Cranford, NJ 07016. Tel.: (201) 272-3600.

Appendix 5

DIRECTORY OF INDIVIDUALS

Abernathy, Prof. Frederick H.: see Harvard University

Allen, Greg: see Allen-Drerup-White Ltd.; also Air Changer Co. Ltd. (The)

Bearg, David W., P.E., 20 Darton St., Concord, MA 01742. Tel.: (617) 369-5680.

Benning, Larry: see Del-Air Systems, Ltd.

Besant, Robert W.: see University of Saskatchewan

Bowser, Dara: see Air Changer Co. Ltd. (The)

Bucher, Axel: see Q-Dot Corp.

Des Champs, Dr. Nicholas H.: see Des Champs Laboratories, Inc.

Dormetta, Ron: see Dravo Corp.

Dumont, Robert S.: see University of Saskatchewan; also National Research Council of Canada

Dutt, Gautam S.: see Princeton University

Eriksson, Krister: see Berner International Corp.

Eyre, David: see Saskatchewan Department of Mineral Resources

Fischer, John: see Berner International Corp.

Fisk, W. J.: see University of California

Fuller, Winslow: see Equinox, Inc.

Giesbrecht, Peter: see Ener-Corp. Management Ltd.

Grimsrud, D. T.: see University of California

Hamilton, Blair: see Memphremagog Group

Holtz, Michael: see United States Government

Jackman, Peter J.: see Air Infiltration Centre

Kays, W. M.: see Stanford University

Kollock, Ray: see Automated Controls & Systems

Kondo, Kyo: see Mitsubishi Electric Corp.

Lefebvre, Ron: see Passive Solar Products Ltd.

Lepage, Arnold: see Memphremagog Group

London, A. Lewis: see Stanford University

Maldonado, E. A. B.: see Iowa State University

Mitchell, Prof. John W.: see University of Wisconsin-Madison

Novecosky, Ed.: see Del-Air Systems, Ltd.

INDEX